HERITAGE, MEMORY AND THE POLITICS OF IDENTITY

T0264650

Heritage, Culture and Identity

Series Editor: Brian Graham,
School of Environmental Sciences, University of Ulster, UK

Other titles in this series

Ireland's Heritages
Critical Perspectives on Memory and Identity
Edited by Mark McCarthy
ISBN 978 0 7546 4012 7

Senses of Place: Senses of Time
Edited by G.J. Ashworth and Brian Graham
ISBN 978 0 7546 4189 6

(Dis)Placing Empire
Renegotiating British Colonial Geographies
Edited by Lindsay J. Proudfoot and Michael M. Roche
ISBN 978 0 7546 4213 8

Preservation, Tourism and Nationalism
The Jewel of the German Past
Joshua Hagen
ISBN 978 0 7546 4324 1

Culture, Urbanism and Planning
Edited by Javier Monclus and Manuel Guardia
ISBN 978 0 7546 4623 5

Tradition, Culture and Development in Africa
Historical Lessons for Modern Development Planning
Ambe J. Njoh
ISBN 978 0 7546 4884 0

Heritage, Memory and the Politics of Identity
New Perspectives on the Cultural Landscape

Edited by

NIAMH MOORE and YVONNE WHELAN

LONDON AND NEW YORK

First published 2007 by Ashgate Publishing

Published 2016 by Routledge
2 Park Square, Milton Park, Abingdon, Oxfordshire OX14 4RN
711 Third Avenue, New York, NY 10017, USA

First issued in paperback 2016

Routledge is an imprint of the Taylor & Francis Group, an informa business

Copyright © Niamh Moore and Yvonne Whelan 2007

Niamh Moore and Yvonne Whelan have asserted their right under the Copyright, Designs and Patents Act, 1988, to be identified as the editors of this work.

All rights reserved. No part of this book may be reprinted or reproduced or utilised in any form or by any electronic, mechanical, or other means, now known or hereafter invented, including photocopying and recording, or in any information storage or retrieval system, without permission in writing from the publishers.

Notice:
Product or corporate names may be trademarks or registered trademarks, and are used only for identification and explanation without intent to infringe.

British Library Cataloguing in Publication Data
Heritage, memory and the politics of identity : new
    perspectives on the cultural landscape
    1. Group identity 2. Collective memory 3. Memorialization
    4. Cultural property
    I. Moore, Niamh II. Whelan, Yvonne
    305.8

Library of Congress Cataloging-in-Publication Data
Heritage, memory and the politics of identity : new perspectives on the cultural landscape / edited by Niamh Moore and Yvonne Whelan.
        p. cm. -- (Heritage, culture and identity)
    Includes index.
    ISBN-13: 978-0-7546-4008-0 1. Art and society. 2. Cultural property.
3. Historic sites. 4. Collective memory. 5. Group identity. I. Moore, Niamh. II. Whelan, Yvonne.

    N72.S6H45 2007
    306.4'7--dc22

                                                                    2006025028

ISBN 13: 978-1-138-24834-2 (pbk)
ISBN 13: 978-0-7546-4008-0 (hbk)

Contents

List of Figures

List of Maps and Tables

Maps

Tables

Notes on Contributors

Mervyn Busteed is a Lecturer in Geography at the University of Manchester.

Paulo Carvalho is an Assistant Professor at the Instituto de Estudos Geográficos, University of Coimbra, Portugal.

Paul Claval is Professor Emeritus, University of Paris-Sorbonne IV.

John Crowley is a Lecturer in Geography at University College Cork.

João Luís Jesus Fernandes is an Assistant Professor at the Instituto de Estudos Geográficos, University of Coimbra, Portugal.

Tim Hall is a Lecturer in Geography at the University of Gloucestershire.

Joanne Maddern is a Lecturer in Geography at the University of Dundee.

Stephen F. Mills is a Senior Lecturer in Heritage and Landscape Studies at Keele University.

Niamh M. Moore is a Lecturer in Geography at University College Dublin.

Tadhg O'Keefe is a Associate Professor of Archaeology at University College Dublin.

Iain Robertson is a Lecturer in Geography at the University of Gloucestershire.

Yvonne Whelan is a Lecturer in Human Geography at the University of Bristol.

Tim Winter is CHASS Postdoctoral Fellow, University of Sydney.

Preface

In recent decades a range of new approaches to the study and interpretation of the cultural landscape have been adopted by cultural, historical and urban geographers, as well as by specialists from disciplines such as archaeology, sociology and architecture. More overtly interpretative and theoretically nuanced approaches to the study of the cultural landscape have evolved as a consequence of the 'cultural turn' across the social sciences and the humanities, resulting in close readings of the signs, symbols and sites of heritage that comprise the landscapes in which we live. Much more than a transparent window through which reality may be unproblematically viewed, the cultural landscape is now conceived of as an emblematic site of representation, a locus of both power and resistance, and a key element in the heritage process. *Heritage, Memory and the Politics of Identity: New Perspectives on the Cultural Landscape* explores the overlapping and oftentimes complex relationships between identity, memory, heritage and the cultural landscape.

The result of a conference jointly organised by the Academy for Irish Cultural Heritages (University of Ulster) and the Department of Geography (now the School of Geography, Planning and Environmental Policy) (UCD), in December 2002, the contributions in this volume interrogate cultural landscapes in both historical and contemporary contexts. The authors are drawn from a range of disciplines including geography, archaeology, and heritage studies and, in common with the other volumes in this series on *Heritage, Culture and Identity*, the geographical scope is global.

In Part I of the book, 'Landscape, Memory and Identity', the contributors focus on the monumental and performative dimensions of memory. Case study examples from Ireland, England, Scotland and the US foreground the significance of the past in the contemporary construction of identity narratives and draw particular attention to the powerful role of monuments and parades as sites of cultural heritage. Tadhg O'Keeffe sets the tone for this part, with a wide-ranging essay that charts the development of the landscape-identity nexus and the politics of collective memory. These themes recur in different guises in the chapters that follow. Iain Robertson and Tim Hall engage with the discourse of heritage in the context of the recent memorialization of acts of land seizure in the Scottish Highlands. Their close reading of the commemoration of conflict on Lewis addresses the intersections between identity, memory and heritage at the local level. In Joanne Maddern's chapter the focus shifts to forms of diasporic identity construction and she explores the discursive representations of immigrant identity and heritage as they are displayed at the Ellis Island Immigration Museum in New York. In particular, she charts the contentious debate and power struggles over memory that prevailed in relation to the erection of a monument to Annie Moore, an immigrant from Ireland and iconic figure in American immigrant culture. John Crowley's chapter focuses on monumental landscapes and in particular upon the ways in which the Great Irish Famine (1845–1852) has been commemorated both in Ireland and overseas. His essay addresses the cultural politics of Famine memory by focusing especially on the role of monuments in Dublin and Boston. In the final chapter in this section, the performance of memory is explored in Mervyn Busteed's richly-detailed analysis of the struggle for ownership of the commemoration rituals for the Manchester martyrs within the Manchester Irish population between 1867 and 1921. He argues that collective memory is central to the formulation of national identity, but that within any group there are alternative readings of the events being commemorated. He demonstrates the ways in which landscape and performance were used for political purposes by successive generations.

In Part II, 'The Politics of Heritage and the Cultural Landscape', the focus shifts to include an examination of the way in which heritage has become politicized for various ends in a range of contexts. The emphasis is on the changing perception of particular heritage sites and buildings, and the role that this has played in constructing and reconstructing particular identities. Paul Claval explores the contemporary transformation in the attitudes of geographers towards landscape and pays particular attention to the role of landscape in cultivating narratives of identity. Drawing on the work of Pierre Nora, Claval explores the changing nature of relations between concepts of place, memory and identity. In 'Valorizing Urban Heritage? Redevelopment in a Changing City', Niamh Moore examines how a particularly important historical building in Dublin's Docklands, Stack A, has become increasingly marketed as a 'flagship' for conservation and urban regeneration. She argues that this is part of a general trend in recent years, whereby economic and planning imperatives have resulted in a heightened awareness of the potential of particular heritage sites. Using the example of this impressive nineteenth century edifice, she highlights how this symbol of industrialization and modernization is being selectively re-imagined and marketed as an icon of post-industrial Dublin. In Chapter 8, Stephen Mills highlights the paradoxical nature of open-air museums in a number of different settings. He contends that the removal of heritage buildings into folk parks alters their symbolic value and creates problematic cultural landscapes. Rather than retaining their unique nature, the identity of these buildings is altered, as they become perceived as representative of a particular type. The final two essays are linked through their emphasis on World Heritage sites. In Chapter 9 Paulo Carvalho and João Luís Jesus Fernandes discuss the town of Elvas, which is applying for World Heritage site classification. In their discussion, the authors examine how the built heritage of this town is being valorized to increase the esteem of the local population, but also to readjust the historical military identity of Elvas as a fortress on the Spanish-Portuguese border. The persistence of linkages between local identity, the planning and development process and the urban landscape are highlighted. In 'Landscapes in the Living Memory: New Year Festivities at Angkor, Cambodia', Tim Winter examines the impact of cultural tourism on the World Heritage site at Angkor, Cambodia. He illustrates the tensions that exist between the promotion of international tourism and the importance of this site to the local people. For Cambodians, Angkor is a symbol of unity and is central to the ongoing reconstruction of a national identity. In contrast, the tourism industry marginalizes such understandings of it as a social, living heritage.

A common thread through each of the contributions is a desire to understand the complex, and often contested, relationships between memory, identity, place, politics and heritage while challenging pre-existing conceptions of their interactions. We hope that the contributions in this volume develop and intensify these debates.

Niamh Moore, Dublin
Yvonne Whelan, Bristol

PART I
Landscape, Memory and Identity

Chapter 1

Landscape and Memory: Historiography, Theory, Methodology

Tadhg O'Keeffe

Introduction

The achievement of the so-called 'cultural turn' more than twenty years ago was to finally secure as orthodoxy the view that the study of society is not objective, ideologically-neutral, value-free, or apolitical, either within or without a scientific method. Liberated by this breakthrough, students of social things and social relationships no longer needed to argue that they were, or are, situated inside rather than outside their field of research, or that society is a construct of humanity rather than a phenomenon of nature, or even that scholarship's intellectual produce is simultaneously and fundamentally social produced. This transformation encouraged many scholars – geographers, anthropologists, archaeologists – outside the narrowly defined and traditionally vocational field of 'social science' to identify a role for themselves in shaping contemporary social agendas, even if there is an argument that no praxis connecting the theory-driven view of the social world to the 'real' world of peoples' experiences and problems was ever developed (Sayer, 2000).

The principal affect of culturalism in the wide humanistic field, shared by the authors in this book, has arguably been a new, albeit often implicit, engagement with the issue of consciousness, as mediated through such counter-Enlightenment – as distinct from post-Enlightenment – projects as phenomenology and hermeneutics (White, 1999). Nowhere is this clearer than in the latest conceptualization of landscape, and its increasingly popular coupling with identity.

The development of a landscape-identity nexus can be followed historiographically. Until the mid-1980s most scholars within the humanities had been happy to treat landscape as a naturally-produced canvas – the metaphor is deliberate – to be primed and painted over by people but to which people are, in a fundamental sense, external. Landscape, in this view, is primordial. It does not require human inhabitation, cognition, or representation to exist. It can be altered by human agency, but it is not, of itself, socially-produced space in any Lefebvrian sense. Thus, most published landscape histories of the decades before the 1980s prefaced their discussions of the cultural features of landscape with descriptions of the natural environment. Braudel's *Mediterranean* (1949), while technically not a landscape history, is a classic example from the *Annaliste* tradition (and indeed one of the first books in which this approach was adopted). Common to all such studies is the identification of a natural, ecologically diverse, landscape on which human activity is inscribed; some of the studies, like Braudel's, attribute or at least imply a strong deterministic role to the primordial landscape.

Although this view of landscape as possessing both natural and cultural layers, with the former being the more deeply-rooted, still informs a lot of landscape-historical research, Marxist-oriented or Marxist-derived work from the mid-1980s to the early 1990s offered an alternative perspective (see Daniels, 1989). Landscape was identified as inherently social-cultural in its production, its cartographic reproduction, and its use, and power was identified as its operating system. It could be argued, though, that by enfolding landscapes into the conspiracies of false consciousness by which élite power is maintained – landscape is implicated in relations of power through its ownership, control and manipulation by social élites – these Marxian readings effectively maintained landscape's externalizing of non-élites, and reduced non-élites' engagements with landscape to acts of compliance (such as the flower-strewn park in front of Kensington Palace after Princess Diana's death in 1997) or resistance (such as the anti-capitalist graffiti that one might find on a public monument during a G8 summit).

Now, however, many of us have bought into the counter-Marxian, and to a large degree constructivist understanding of 'landscape' that took root during the 1990s (Mahoney, 2004). The critical change is that, whereas the Marxian view of landscape-as-power tended to externalize (and victimize?) non-élites, among whom we number ourselves, we now see ourselves and others as situated inside landscapes, forming and reforming them. On a simple level, the funeral vista and graffitied wall can now be understood not as landscape-situated responses (of compliance, of resistance) to authority that is articulated in the landscape, but as acts of landscape-construction and so of identity-formation in their own right. More abstractly, we claim landscapes to be 'spaces' or 'places', or both simultaneously, that exist reflexively in our cognitive as well as our corporeal experiences of the material world, shaping and being shaped by our simultaneously multiple identities as humans. Landscape, then, is now characterized implicitly as a product of mindscape, to borrow a word from Zerubavel (1997). Its connection with the realms of the cognitive and mnemonic, and so with the general issue of consciousness (including 'non-consciousness', in the sense of Bourdieu's 'habitus'), is therefore inalienable. So too is its democratic value: everybody knows, possesses and partakes in 'landscape'. Here it intersects with Raphael Samuel's 'resurrectionism', a model of history that embraces the popular and vernacular, the feminine and domestic, and is 'inconceivably more democratic than the earlier [models]' (Samuel, 1994, 160).

History and the Politics of Collective Memory

The relationship of history to memory has long been a central issue in epistemological debates within the historical sciences. Without getting bogged down in its historiography, there is at the heart of the debate the sort of confusion about terminology and meaning that is often generated when topics from one discipline are dragged across the boundaries of another. The view of the relationship to which I subscribe is that history, which is narratological, is always about memory, the first implication being that memory is larger, or something more, than history, and the second being that history cannot ever claim to be any more than one line, or one cluster of lines, bringing the past into the present.

The definition of history raises a range of issues, and one hesitates to allow the academic discipline of History complete custody of what the term constitutes, but we can probably agree, first, that a document-form record or testimony that we customarily describe today as historical is an artefact that contains something which its author wished to be remembered for some duration, however brief, and secondly, that historians reconstruct pasts by sculpting idealised collective (after Halbwachs, 1950) or, better still, collected (after Young, 1993) memories out of such raw 'historical' material. 'Historical memory' can be regarded, therefore, as that of which we are reminded, as distinct from that which we remember, and 'historical consciousness' can be regarded as alluding to that part of the mindscape that chooses history, or rather historicity, as the model for ordering the world of past experience. To my mind, then, collected memory is always historical (or narratological) and is always the product of some programme of being-reminded. However, and at whatever scale a collective is constituted, we have no collective capacity to share memories that are not in some way externally programmed for us.

For example, I, like so many others, have memories of the visit of Pope John Paul II to Ireland in 1979 and the day-long mass in Dublin's Phoenix Park. On the one hand, I have memories that are intensely personal since they relate to my sensual (mild claustrophobia) and emotional (teenage melancholy!) experiences of the time. To properly visit those memories now, sitting in my study in Dublin, would actually involve great mental energy. Empathetic remembering demands that one 'switches off' (or, to use computer language, 'quits other applications') in order to create the appropriate memory space. So, I could reconnect to some extent with those sensual and emotional experiences right now, but I would need to disconnect from what I am doing – writing this paper, listening to music but also hearing the background silence should it be broken by the baby crying – in order to concentrate on doing so. On the other hand, I have other memories of that day in the park that owe more to the mediation of the event (through television, for example, see Bourdon, 2003) than to my personal experience of it. Those memories are visual-factual rather than sensual-emotional. I could recount them here and now and still be able to listen to music in the background. These are the memories of mine that I regard as commonly possessed, as elements of a collected memory. In fact, they are not really memories of the event but memories of its mediation. I do not actually remember a congregation of more than one million people because I saw no more than several thousand, so my memory of the vast gathering of people is a memory of its image.

I argue, using this simple personal example, that collected memory is a product of external programming, and that – borrowing words used by Gerald Vizenor (1998) in a different though not-unrelated context – it represents a triumph of 'simulation' over 'trace'. Before developing that point here with respect to landscape, I suggest that there is a moral imperative to accept this proposition about collected memory: the opposite concept, which is of an intuitive collective memory, is dangerously essentialist, since it burdens the individual with a store of memory over which he or she has no control, and potentially ensnares the individual in a web of collective responsibility.

Landscape is obviously a touchstone for remembering both the visual-factual and the sensual-emotional. This function is not a by-product of landscape but is integral to its definition. While the new landscape culturalism emphasizes our experiential and empathetic engagements with the world, and suggests a study of landscape mynomony that acknowledges the importance of sound and smell, most scholars continue to privilege vision over other senses. The study of landscape and memory often devolves, therefore, into a study of tangible, visual, *aides de mémoire* within landscapes. This is certainly the case with respect to western capitalist societies, where its origin can be traced back to the Renaissance 'theatre of memory' and further back into classical times (Yates, 1978). The implication that the capacity to 'read' ancestral memory and locate identity in the non-monumentalized landscapes is the preserve of indigenous non-western peoples (as documented by Morphy, 1995 and Santos-Granero, 1998, for example) is a troubling one.

Lieux de Mémoire

A number of scholars have mapped the ways in which personal memories have been reshaped into collective memories by forms of political intervention in western capitalist landscapes, particularly through 'official' acts and objects of commemoration (White, 1999; Shackel, 2001). Among the most interesting explorations are those marshalled by *Annaliste* scholar Pierre Nora in the multi-volume study of *lieux de mémoire* in France (Nora 1997, 1998, 2001). Placing memory at the centre of the French psyche, he contends inter alia that the cohesion of 'France' as both an object and a locus of national identity is preserved in a range of memorial forms devised especially by state institutions and encountered as places of memory in people's individual daily and public rituals. Nora laments what he sees as a decline of a national, collective, identity-forming, memory in the age of globalization, just as Hough (1990), operating at regional rather than national scales, sees market forces and technologization as agents that homogenize the landscapes of places that were otherwise unique, thus depriving them of their special character (see also Zukin, 1991). Nora even attributes its 'demolition' to the 'terrorism of historicized memory' as promoted by the academic discipline of History (Nora, 1989, 14).

Nora draws our attention to the importance of rupture in the generation of collected memory, the point being that moments of social stress or fracture ignite

Figure 1.1   Remembering War at the Cenotaph, Belfast, November 2004
Source: Tadhg O'Keeffe

desires to collect memories that can be shared. Fritzsche (2004) has shown how, for example, in places like New England, the rupturing effect of nineteenth-century 'progress' generated a nostalgic, melancholic, yearning for a forgotten past that, in turn, generated conscious strategies of memorializing, not least within the domestic landscape of the household, itself a place of collectivity. Nora himself attributed changes in the collective mnemonic to the cataclysmic ruptures of war. He saw the French Revolution as a critical moment in the crystallizing of a French national memory embracing France's medieval past, and saw twentieth-century world wars, especially the second, as critical to that shift from memory to history that he laments. It could be argued, though, that the great wars significantly enlarged the field of collected memory, generating new rituals of public commemoration on the one hand (Figure 1.1) and, as Samuel has argued, a new tradition of regarding the *bric-a-brac* of everyday, vernacular, contemporary culture as heritage on the other (Samuel, 1994).

The combining of rupture, as discussed by Nora, and vernacularization, as discussed by Samuel, to reassert national identity finds a dramatic exemplification in two elements of the story of the World Trade Centre after its violent destruction on September 11, 2001. First, those everyday, vernacular objects, such as briefcases, that were retrieved from the naked, smouldering core of downtown Manhattan acquired from their context of rupture a magnitude of meaning as heritage objects that would have been quite unimaginable in other circumstances. They were asked, in the very title of the Smithsonian Institute exhibition in which they found a home, to 'bear witness to history'. Second, as modernist works, the World Trade Centre towers had denied historicity during their lifetimes – they celebrated the materials of their present (glass and steel) and insisted that their aesthetic was their structural honesty, and they presented a Miesian environment of unadorned structural transparency to an industry (bond trading/non-retail banking) that only deals with the present and the future. But as ruins they acquired historicity, partly by virtue of being ruins, since ruins imply 'history', and partly by recalling those French Gothic buildings that, in their own medieval age, represented modernity. So now, having once possessed a modernism that refused to incorporate memory, the towers find themselves the objects of a memory cult.

Pierre Nora's multi-volume project was fundamentally political: if the bonding agents of French national identity are ever stripped away by some conspiratorial alliance of History, as Nora understands it, and globalization, the volumes will fill the space that is left, becoming a rallying point for a reinvigorated national cultural consciousness. *Lieux de mémoire* will thus become a *lieu de mémoire* in its own right. I suggest that there is an interesting parallel here with *The Atlas of the Irish Rural Landscape* (Aalen et al, 1997), a work very familiar to Irish readers of this book. This sumptuously-produced hardback volume of images and maps, with accompanying text, probably belongs in more households in Ireland than any other hardback of its size and substance. It is a cousin of the famous technicolour postcards of Ireland produced by John Hinde from 1957 to 1972. Those Hinde postcards celebrated Ireland's rurality on the one hand, and, with images of holiday camps and Art Deco airport terminals, saluted Ireland's modernity – embryonic modernity to be sure – on the other. We posted to all corners of the world our achievement in retaining diagnostic rural traditions in their beautiful settings while managing to be part of the larger world. The Atlas – is it really an atlas, and what are the connotations of that word? – also celebrates the rural and acknowledges that the future holds challenges. Yet its focus on the rural (when most Irish people live in cities), its 'man and his habitat' language, its untheorized understanding of landscape, its under-representation of the rural landscapes of Unionist communities, its lack of women, and so on, all suggest that it is no less a project of an 'invented Ireland' than the John Hinde postcards forty years earlier. Indeed, the appearance of this volume at the height of the Celtic Tiger era suggests that, like Nora's great compilation, it captures and memorializes its country's innate cultural profile at the moment of globalized threat. Like *Lieux de mémoire* the Atlas is a *lieu de mémoire* in its own right.

Methodology and Representation: Heritage and Landscape Management

The rather radical change in thinking about landscape that I introduced at the start can be understood as an importation into landscape research of culturalism's insistence that social formations (such as identity) and social institutions (such as 'the market') are fluid and contingent, rather than primordial, cross-cultural and transhistorical. One good indicator of the most recent shift in thinking is Mitchell's preface to the enlarged 2002 edition of *Landscape and Power*, originally published in 1994, in which he acknowledges that *Space, Place, and Landscape* would now be a better title (Mitchell, 2002). The longer-term shift from one way of thinking to another is easily gauged by noting how terms like 'paradigm', 'pattern', 'model', to select a few at random from the era of New Geography and New Archaeology, have now largely disappeared from the landscape-themed literature emanating from the humanities, while terms like 'negotiation', 'embodiment', 'selfhood', 'performance', 'memory', are everywhere. Where once it was fashionable to use logarithms to test hypotheses about distributions within landscapes, it is now fashionable to use landscapes to do such things as 'negotiate identity at remembered boundaries of gendered selfhood', whatever such things mean; that is not a quote, by the way, but it may transpire to be an inadvertent one!

This is an exaggerated polarization, of course: the paradigmatic shift – let's continue to use that phrase – between what we might describe as modernist landscape scientism and post-modernist landscape culturalism has actually not been as dramatic, nor even as universally adopted, as first seems to be the case, and one could argue that less extreme versions of landscape culturalism have been around for a long time. Indeed, one might even argue that the landscape on which the postmodernist's gaze is fixed and the landscape on which everybody else's gaze is fixed are actually very different things, and that the use of the same term – landscape – complicates matters needlessly. But the polarization serves a purpose in the context of this essay, since it defines extremes between which are the positions that we should surely seek to occupy.

The primary and on-going challenge for scholars who generally embraced elements of the post-modernism that emerged in the early 1980s is to find methodologies that locate the middle ground where the rigour of scientism can be combined with the imaginative freedom of culturalism to produce meaningful, non-relativist, and indeed politically useful (an idea I have adapted from Wainwright, 2000) understandings of landscape. There is also a challenge of language. If the new concept of landscape is one of democracy, so too must the language of the new 'landscapism'. A critical language (in the conventional sense) that is impenetrable and obscurantist, as some postmodern writing is, excludes rather than includes, and regenerates the concept of power – academic/intellectual power, in this case – within a field in which the concept of power has lost much of its potency. Language is also important on a theoretical level: as Peter Carruthers has argued in his study of consciousness, language, or rather linguistic structure, is critical to reflexive cognition itself critical to the new culturalist characterization of landscape (Carruthers, 1996).

These are intellectual challenges: for example, how do we, or should we even attempt to, reconcile the ways of thinking in which we have been either culturally inculcated or academically trained with the sorts of outcome that we aspire to achieve? But they are also challenges that extend into the world of landscape management. Sophisticated management requires, after all, a sophisticated understanding of landscape's many levels of operation. Natural ecology falls outside our remit – although it is interesting to note in passing that the category 'natural landscape' is a casualty of new thinking, since the very concept of nature is now claimed to be culturally constructed (Shepheard, 1997) – but the social role of landscape does not. A core belief of landscape culturalism is that communities invest in landscape formation, and that they locate their identities within landscapes or have their identities metaphorized as landscapes. Thus all landscapes, from the bleakest urban industrial wastelands to the most verdant country estates, are social-ecological, and all landscapes qualify as somebody's heritage.

Whose Heritage?

Before reflecting in the final two sections on the implication of this for landscape management, it is important to state at the outset that the concept of heritage as currently constituted, and as currently curated by a self-styled industry, is more problematic than helpful. The 'heritage industry' has its own menu of criteria by which landscapes can be evaluated as cultural-historical 'things', especially for preservation or non-preservation in the face of demands from other, more environmentally aggressive, types of industry. While these criteria – antiquity, uniqueness, history and/or historical association – may seem reasonable, ideologies other than that ideology of democracy, which I mentioned above, kick in when landscapes are selected according to these criteria. This is especially apparent in urban environments: for example, those districts in American cities that are characterized as 'historic' are often occupied by non-white communities, often in 'transient' architecture, and yet their histories are not remembered in official heritage discourse (Hamer, 1998). Closer to home, the concept of heritage in Dublin, a city with a strong sense of its own past, is not extended to its immigrant communities. Yes, these communities are relatively recent arrivals but they are here to stay, and yet the earliest sites of, say, African occupation – places to be revered by later generations of Afro-Irish – have been afforded no protection by the state. Indeed, as I write, the fate of a house on Dublin's Moore Street is being debated. The area around this street is being redeveloped, and its buildings are earmarked for demolition. The derelict house in question (Figure 1.2) was the last refuge and site of surrender of the rebels who occupied the General Post Office during Dublin's heavily-mythologized 'uprising' of 1916. The likelihood is that the house will be saved, not for its architectural merits (although it does possess some) but for its great historical significance. The indications that it will be saved are welcome, even if the fact of its dereliction makes one wonder why it was not worthy of conservation long before its survival was threatened. Observe that house's location in a streetscape, however, and we can reasonably claim that the impending demolition of the other nineteenth-century houses reflects an undervaluation of Dublin's working class heritage from the last two centuries and a disregard for the future heritages of Ireland's immigrant communities. Heritage tends to be white, yet it need not be so (see Samuel, 1994).

Now, my argument is that all landscapes qualify as somebody's heritage, even if the term 'heritage' is rarely allowed such liberal application. This is certainly not an argument for the preservation of all landscapes, almost as adjuncts to peoples' civil rights; quite aside from the absurdity of such a proposition, the concept of a preserved landscape is clearly a contradiction in terms, since a landscape that is artificially sealed at a particular moment stops being the landscape that it was and becomes a new landscape. But it is an argument for some cognisance among landscape managers of how identities, many of which are historically rooted, are actually inscribed in landscapes, even in ones that seem very mundane.

Figure 1.2   Moore Street, Dublin, November 2005
The 1916 house is second from the right. It is flanked by houses now converted to business premises for some of Dublin's immigrant communities, and a Nigerian Pentecostal church occupies the rear of one of them. Some of the market stalls of 'indigenous' street traders are visible in the foreground.
Source: Tadhg O'Keeffe

And so we come back to the question of methodology and language/ representation. How can we capture those identities through fieldwork that is both within and on landscapes? How can we advocate some accommodation of identity-heritage in the discourses and strategies of landscape management? And what particular accommodation of them would we seek? These are questions that cannot be answered here, but it is important to ask them as a first step.

Building V(o)ices: Introducing the Landscape of Dublin's Monto

I wish, finally, to introduce a case-study that draws together a few of these ideas about landscape, memory and heritage. The case-study is again in Dublin's north inner city: it is the area that was known in the nineteenth century as 'the Monto', a name that was an abbreviation of Montgomery Street, one of its main thoroughfares (Finegan, 1978). A series of late Georgian and early Victorian terraced houses in a small area halfway between the Custom House and Mountjoy Square, it was a red-light and (often-illicit) alcohol-retailing district of considerable notoriety for most of the 1800s and the early 1900s. James Joyce christened it Nightown, a name that captured well the furtive nature of its best-known industries. It seems to have had established boundaries: for example, the opening lines of Chapter 15 of *Ulysses*, the chapter that is set within it, introduce us via one of its entrances to its Dickensian fog, its flickering lights, and its night-sounds.

Although prostitutes of the Monto often had high-profile clients (Fagan, n.d., 19), it was probably not a place into which many middle-class people casually wandered during the late 1800s and early 1900s, particularly in the evenings. The menace of drunken men and soliciting women may have repelled the mildly curious. Anyway, crossing the boundary into the area was probably a great social challenge: a man might not meet his wife in there, but she might see him going in. The realization that cameras of moral surveillance were probably trained on its entrances may have discouraged many a prospective middle-class visitor. The Monto's customers were mainly working-class Dubliners, as well as British soldiers stationed in Aldborough Barracks nearby, and sailors newly arrived in Dublin port, a short walk to the south.

However, areas such as the Monto, and Little Lon in Melbourne, to name a comparable nineteenth-century urban red-light district elsewhere, did attract social reformers, educators, and missionaries. By the start of the twentieth century there was a significant religious presence in the Monto, ranging from convents with laundries, where clothes and souls were washed together, to so-called midnight missions, which by their very terminology emphasized the area's nocturnalism. Indeed, it was the intervention of Frank Duff, founder of the Roman Catholic charitable organization, the Legion of Mary, which eventually brought about the demise of the Monto. In the early 1920s, Duff had witnessed the suffering of prostitutes within the area and he successfully mobilized his organization against the madams. Thanks mainly to his initiative, the brothels closed down shortly after the foundation of the Free State, and in the following years the buildings were demolished and replaced by public housing ('flats'), and priests from Dublin's Catholic Pro-Cathedral blessed the area. Virtually nothing remains in the modern landscape of the Monto's red-light fabric, and even the original nineteenth-century street names are gone: Montgomery, Mecklenburgh, and Mabbot, street-names of genuine resonance in the Dublin of the later 1800s, became Foley, Railway, and Corporation, and Corporation Street has now been changed again to James Joyce Street.

Returning to the issue of how we conceptualize landscape intellectually, one could certainly conceive of the Monto landscape as one that was encoded by and for power relations, and was thus a theatre in which strategies of contestation and resistance, built around issues of gender and commodification, were played out. Given how much we know about the early twentieth century in the Monto, this is an obvious approach to reading its now-lost landscape. Our lack of detailed information about the actual, up-close, physical appearance of the area – remember, much of it was cleared away between seventy and eighty years ago with no proper record made of it – does make it difficult to apply this power-resistance model in anything other than broad brush-strokes, but that is not the point. However, the problem with characterizing the Monto as a place of power and resistance is that it essentializes its categories of occupant, so that the madams, 'clients' (as they might be described in contemporary language), prostitutes, reformers, and others, are characterized as the occupants of separately-defined niches within the power-play. Why is this a problem? It is because essentialism is a dangerous concept. It denies agency to individual social actors. As human scientists we instinctively know social relations to be fluid (contingent) rather than fixed (essential), so we would have no difficulty imagining, for example, very complex, sometimes contradictory, social interplays between the parties of the Monto. This is the anti-essentialist perspective.

Let us think for a moment about what that means in terms of interpreting prostitution in the area. One could not deny for a moment the abject misery of life 'on the street' for the Monto's prostitutes. It is an historical fact that so many of them experienced violence, were alcoholics, and contracted (and often died as a result of) venereal disease. But one could argue against defining these girls' lives in the Monto solely in terms of their prostitution. Prostitutes are, after all, social actors both inside and outside the context of prostitution, even if their agency is sometimes severely compromised by the affects of alcohol (as in the Monto) and drugs (as in modern Dublin) on their cognition of the world, and therefore on their self-awareness. Moreover, if we focus exclusively on the suffering experienced by the Monto prostitutes, real though that was, we might not recognize just how important were the prostitutes as nodal points in the spatial and cognitive geographies of social relationships. The history of the Monto, then, is not just about the commercialization of the body under duress, but also about networks of social relationships. So, while the popular perception of the end-game of the Monto in the 1920s – Frank Duff leads a posse of legionnaires through its streets and defeats the forces of immorality in a straight shoot-out – certainly appeals to our sense of the triumph of good over evil, the anti-essentialist position would seek a partial explanation for the demise of the Monto in its complex social relations, many of which used prostitution as a conduit.

That very point brings us back to the alternative, culturalist, characterization of the Monto as multiple imbricated landscapes, perceived and negotiated side-by-side and interactively by the individual madams, policemen, ramblers (see Rendell, 2002), drunkards, prostitutes, and others, as well as by collectives of these. Retrieving these landscapes is virtually impossible, again because of a paucity of information, but a project that I am currently engaged in – the reconstruction on paper of the Monto street elevations as they were in Joyce's era – may provide some illumination by virtue of retrieving something of that landscape's appearance.

Remembering and Forgetting the Monto

Earlier in this chapter I asked three linked questions about capturing and managing the multiple identities that, according to the culturalist perspective, are invested in landscapes. I want to finish here by commenting on how we rise to those challenges

**Figure 1.3   Railway Street (formerly Mecklenburgh Street), Dublin,
                 December 2005**

The pair of training shoes dangling above Railway Street signifies that heroin is on sale here.
Source: Tadhg O'Keeffe

in the area that was the Monto, aware of the fact that no general strategy or manifesto
can emerge from any single case study.

The Monto no longer exists in any physical sense. There is nothing in the area
today to remind one of its past. It is no longer the area of prostitution that it was, drugs
have replaced alcohol as a principal substance of abuse, and training shoes draped
across over-head lines (Figure 1.3) have replaced flickering candles in fan-lights as
the signifiers that illicit product is on sale, and modern apartments are now beginning
to alter the area's social landscape. The recent rebranding of Mabbot/Corporation
Street as James Joyce Street might suggest at first glance that Dublin's municipal
authorities are now inviting us to remember Nightown, but in fact it is Joyce who is
being commemorated, and few passers-by looking up at the street name will realize
that here was an entrance to the Monto (Figure 1.4). We do at least have the recorded
memories of residents of the area to help us imagine the Monto, and these, combined
with photographs, provide a vivid testimony to the character, and the characters, of the
area (Fagan, n.d.). Inevitably, however much the testimonies suggest the possibility
of multiple narratives about the Monto, they are understood to be fragments of a
single story, as witness, for example, the sub-title of the fine book on the Monto by

Figure 1.4   Mabbot Lane, Dublin
The name Mabbot now survives only in a small laneway to the west of James Joyce Street.
Source: Tadhg O'Keeffe

Terry Fagan and the North Inner City Folklore Project (n.d.). These testimonies now constitute an historic source in themselves, since there are relatively few people alive today, anywhere within the entire city of Dublin, who can recall with any clarity that heyday of the madams that ended more than eight decades ago.

Some of the published recollections have, to my mind, the strong scent of nostalgia, a yearning for 'rare 'oul times' among the long-gone tight-knit communities of long-lost streets. This is not to query their honesty or accuracy. Nor is it to suggest that the residents of the area who were not involved themselves in the Monto's two industries were indifferent to the plight of prostitutes within their landscape and community. On the contrary, sincere compassion is present in virtually all the recorded memories, and there is no doubt that the success of Frank Duff's mission owed much to that compassion. But one wonders if the post-Monto history of the north inner city, with its demolition of the historic properties after 1925, the various schemes of public housing that replaced them over the decades, and the chronic drug epidemic that visited the area in 1970s and 1980s, all ruptures of one sort or another, affected memories of the Monto's closing years. We know what has been remembered, and much of it is anecdotal, but what has been forgotten? Ironically, Joyce's fictive Nightown suggests what a sensual as distinct from an anecdotal recollection might comprise.

People who were born there and spent most of their lives there formulated the published recollections inside the Monto. Indeed, there is in many of these recollections a sense that the Monto was 'an inside', a place on which there was an insider's perspective, even if it is not clear from these recollections what the Monto was regarded or perceived as being 'inside' of. It was, of course, a geographical area inside the city, but the relationship of it to the larger urban area was more than simply spatial. That sense of it being a place apart is best represented by those inversions of convention within the Monto to which Gary Boyd (2002) has drawn attention, such as the way in which its main business thoroughfares were actually its back-streets, or the ways in which the madams exercized control inside it whereas men exercized control in the city outside it. Even its characterization as Nightown is an inversion.

We can reconceptualize these inversions as evidence that the Monto was simultaneously 'an outside'. In a spatial sense it was as much outside the city as within, as its network of streets was developed initially in a small wedge of land – the Aldborough estate, to be specific – hemmed in between the city's dockland and one of its great Georgian developments. More significantly, the normal etiquette of urban life stopped at its boundaries, as if this were an extra-mural settlement. Men moved outside their normal lives, and sometimes outside respectability, when they rambled through it or sought sex within it. We know the girls themselves to have been outsiders, from elsewhere in Dublin or from the country.

The general absence of historic fabric raises questions about the invoking of history within the context of debates on the area's redevelopment. It means that any strategy of physical redevelopment that seeks to accommodate the area's past will involve 'simulation' rather than 'trace'. There is nothing there that is a trace; collected memory, for what it's worth, is not rooted to the landscape that we see today. The questions, then, are: what is to be simulated, in what way, and for whom? They are questions I cannot answer.

Coda

As the old Monto undergoes slow rejuvenation, one is conscious today of what has happened around its fringes. The area to the north-west is now a commercial sector for Dublin's new working class: its immigrant community. Most of the buildings are old, and many of them were built around the same time as the lost houses of the Monto, but now they include Internet shops and call-centres among their retail functions. Thus they reflect the connectedness of this district and of this working class community to a wider world. At the other end of the old Monto is the Custom House, the International Financial Services Centre, and Busáras, the bus station. The first two of these, built almost exactly two centuries apart, also connect to a wider world, one historically and one contemporaneously. Busáras, to my mind, connects, or rather connected, to something quite different. It connected to a future that never transpired. A masterpiece of the post-war international architectural style, the building of Busáras in the landscape of 1950s Dublin suggested through its monumental incongruity a metaphoric pathway into modernity for the Dublin of the second half of the twentieth century. But it was a pathway that was never followed. The building remained a relatively isolated venture as the city retrenched, architecturally and culturally, in the 1960s. While the Celtic Tiger brought some sophistication many years later, Dublin's built landscape has barely dragged itself to the level that Busáras suggested was possible: looking at the proliferation of clocks that appeared on buildings of the 1990s in Dublin makes the point very well: the city's architecture during the boom years did not so much play with time, as postmodernism does, as simply tell the time.

References

Aalen, F.H.A., Whelan, K. and Stout, M. (eds) (1997), *The Atlas of the Irish Rural Landscape*, Cork University Press, Cork.

Bourdon, J. (2003), 'Some Sense of Time: Remembering Television', *History and Memory*, 15, 2, pp. 5–35.

Boyd, G. (2002), 'Legitimising the Illicit. Dublin's Temple Bar and the Monto', *Tracings*, 2, pp. 112–25.

Braudel, F. (1949), *La Méditerranée et le Monde Méditerranéen à l'Epoque de Philippe II*, Armand Colin, Paris.

Carruthers, P. (1996), *Language, Thought and Consciousness: An Essay in Philosophical Psychology*, Cambridge University Press, Cambridge.

Daniels, S.J. (1989), 'Marxism, Culture and the Duplicity of Landscape', in Peet, R. and Thrift, N. (eds), *New Models in Geography: The Political Economy Perspective*, Unwin Hyman, London, pp. 196–220.

Fagan, P. (no date), *Monto: The Story Behind Dublin's Notorious Red-Light District As Told By The People Who Lived There*, North Inner City Folklore Project, Dublin.

Finegan, J. (1978), *The Story of the Monto*, Mercier Press, Cork.

Fritzsche, P. (2004), *Stranded in the Present: Modern Time and the Melancholy of History*, Harvard University Press, Cambridge Mass.

Halbwachs, M. (1950), *The Collective Memory*, Harper and Row, New York.

Hamer, D. (1998), *History in Urban Places: the Historic Districts of the United States*, Ohio State University Press, Columbus.

Hough, M. (1990), *Out of Place: Restoring Identity to the Regional Landscape*, Yale University Press, New Haven.

Mahoney, M.J. (2004), 'What is Constructivism and Why is it Growing?', *Contemporary Psychology*, 49, pp. 360–363.

Mitchell, W.J.T. (ed.) (2002), *Landscape and Power*, University of Chicago Press, Chicago.

Morphy, H. (1995), 'Landscape and the Reproduction of the Ancestral Past', in Hirsch, E. and O'Hanlon, M. (eds), *The Anthropology of Landscape: Perspectives on Place and Space*, Clarendon Press, Oxford, pp. 184–209.

Nora, P. (1989), 'Between Memory and History: *Les Lieux de Mémoire*', *Representations*, 26, pp. 7–25.

Nora, P. (ed.) (1997), *Realms of Memory: The Construction of the French Past, II: Traditions*, Columbia University Press, New York.

Nora, P. (ed.) (1998), *Realms of Memory: The Construction of the French Past, III: The Symbols*, Columbia University Press, New York.

Nora, P. (ed.) (2001), *Rethinking France: Les Lieux de Mémoire, I: The State*, University of Chicago Press, Chicago.

Rendell, J. (2002), *The Pursuit of Pleasure: Gender, Space and Architecture in Regency London*, Athlone Press, London.

Samuel, R. (1994), *Theatres of Memory*, Verso, London.

Santos-Granero, F. (1998), 'Writing History into the Landscape: Space, Myth and Ritual in Contemporary Amazonia', *American Ethnologist*, 25, 2, pp. 128–48.

Sayer, A. (2000), 'Critical and Uncritical Turns', in Naylor, S., Ryan, J., Cook, I. and Crouch, D. (eds), *Cultural Turns/Geographical Turns: Perspectives on Cultural Geography*, Prentice Hall, Harlow, pp. 166–81.

Shackel, P.A. (2001), 'Public Memory and the Search for Power in American Historical Archaeology', *American Anthropologist*, 103, 3, pp. 655–670.

Shepheard, P. (1997), *The Cultivated Wilderness: or, What is Landscape?*, MIT Press, Cambridge, Mass.

Vizenor, G. (1998), *Fugitive Poses. Native American Scenes of Absence and Presence*, University of Nebraska Press, Lincoln NE.

Wainwright, H. (2000), 'Political Frustrations in the Post-Modern Fog', in Philo, G. and Miller, D. (eds), *Market Killing: What the Free Market Does and What Social Scientists Can Do About It*, Longman, Harlow.

White, G.M. (1999), 'Emotional Remembering: The Pragmatics of National Memory', *Ethos*, 27, 4, pp. 505–29.

White, H. (1999), 'Afterword', in Bonnell, V.E. and Hunt, L. (eds), *Beyond the Cultural Turn: New Directions in the Study of Society and Culture*, University of California Press, Berkeley, pp. 315–24.

Yates, F.A. (1978), *The Art of Memory*, Penguin, Harmondsworth.

Young, J. E. (1993), *The Texture of Memory. Holocaust, Memorials and Meaning*, Yale University Press, New Haven.

Zerubavel, E. (1997), *Social Mindscapes. An Invitation to Cognitive Sociology*, Harvard University Press, Cambridge Mass.

Zukin, S. (1991), *Landscapes of Power: From Detroit to Disney World*, University of California Press, Berkeley.

Chapter 2

Memory, Identity and the Memorialization of Conflict in the Scottish Highlands

Iain Robertson[1] and Tim Hall

> Where memory is no longer everywhere, it will not be anywhere unless one takes the responsibility to recapture it through individual means. (Nora, 1986, 6)

> The prime function of memory ... is not to preserve the past but to adapt it so as to enrich and manipulate the present. (Lowenthal, 1985, 210)

When we imagine Scotland it is, with one or two notable exceptions, the Highland heritage landscape that comes to mind, with its received identity as a land of loch, glen and mist-covered mountains; a place of whisky, shortbread and tartan and a space that is idyllic, open and empty (Gold and Gold, 1995; McCrone et al, 1995; Withers, 1992). If people appear at all in this myth, then they are a happy carefree peasantry, living an Arcadian existence. This idyllic representation fails to account for the socio-economic transformation of Highland society and the concomitant system of class exploitation that emptied the region in the first place and which is recorded in the Highland heritage landscape of the present day.[2] This chapter explores the politics of representation in the Scottish Highlands by focussing on the erection of a series of memorial cairns to land disturbances on the island of Lewis in the Outer Hebrides in the mid-1990s. In that they create and celebrate a landscape of resistance to the social and cultural processes that underlay the emptying of the Highlands and permitted the making of the Highland heritage landscape myth, this chapter will

---

1 Iain Robertson would like to dedicate this chapter to the memory, in both senses of the word, of Angus 'Ease' Macleod. Angus was not only chair and driving force behind the process of heritage making discussed here, but he also had a deep knowledge of the social and cultural history of the Lewis crofting community and of Pairc in particular. Angus knew many of the participants in the last manifestation of the land disturbances and I will always be grateful for his generosity in sharing those memories in conversation and in many (very long!) letters.

2 The best, recent, survey of the complex changes in Highland society in this period is Withers, 1989. See also Gray, 1957, pp. 57–141; Macinnes, 1996; Dodgshon, 1998. Tartantry has been much discussed in the academic literature, see McCrone, Morris and Kiely, 1995, pp. 49–61; Harper and Vance, 1999, pp. 35–6.

argue that the Lewis memorial cairns are a creation and celebration of heritage from below, and an antidote to the otherwise widespread process of myth-making.

Social and cultural change in the northwest Highland and islands took the form of the replacement of small-scale subsistence agriculture with large-scale sheep farming. It also involved the transformation of part feudal, part tribal social relations into purely capitalist relations, and the concomitant transformation of clan chiefs into landlords. In the period from about 1820 to 1850 it involved the forced migration and emigration of the bulk of the Highland population in a process that has become known as the Highland Clearances. Highland communities did not accept change passively and, over time, their reaction took a number of forms. Resistance to clearance was not universally apparent (partly a consequence of Highlanders' retention of their feudal/tribal *mentalité*) although there were some attempts. In the 1880s, however, a more confident surviving crofting population sought to reassert what they saw as their lost rights to land (a belief that drew on the same *mentalité*). One of the chief features of this period of land disturbance, known as the Highland Land Wars, was the land raid: the forced seizure of specific but often quite small parcels of land. The raiders believed that cultivation of the land by their forebears gave them rights of occupation and utilisation. Disturbance was sustained into the 1890s and was revived after World War One.

Both the Clearances and the Land Wars have been memorialized in the landscape in recent years and both sets of commemorations have received attention in the academic literature (Basu, 1997; Withers, 1989). Much of this research has focussed on the constitutive role of memorials in identity formation and contestation. More generally, literature on the relationship between heritage and identity and the role of memorialization in that relationship has focussed chiefly on expressions of nationalism (or supra-nationalism) and the deployment of narratives of the past to construct contestable and contested geographies of belonging in the present. It is the contention of this chapter, however, that local manifestations of identity making are no less fractured and contested and are of equal significance for the individuals and groups concerned. We argue, therefore, that attention should be paid to local identity and to the ways in which it is mediated through contested narratives of the past. This chapter focuses on a very localized and specific manifestation of the deployment of the past in the contemporary manufacture and maintenance of identity narratives. While rather traditional and vernacular memorials to the Land Wars were first erected on the islands of Tiree and Skye, the monuments that are the subject of this chapter eschew the vernacular in favour of the representational in an attempt to create emblematic landscapes of resistance. They are also expressive of a powerful relationship between memory, memorialization and the making of heritage.

Memory, Memorialization and the Making of Heritage

The process of remembering and of identity and heritage creation is neither autonomous nor uni-directional. Heritage, identity and cultural landscapes draw on discursive practices that are, of themselves, complex and contradictory, so we must view such landscape representations as similarly capable of generating contestation and conflict. There is, then, a malleability to heritage and its relationship to landscape, that derives in part from the dialectical relationship between memory and history.

Two of the most significant commentaries on this relationship have been those by Halbwachs (1992) and Nora (1989) and there appears to be much congruity between them. Both argue that memory is collective, plural and yet also individual. At a basic level there is a shared belief that modern memory needs to be sustained through mnemonics. We can recognize similarities in their explication of the origins of this dialectical relationship: a transition between one form of memory and another; but, critically, profound dissimilarity in their understanding of the nature of this transition. Where, for Nora the process is viewed as generating a distinct discontinuity, Halbwachs is concerned to recognize a near benign shift from 'historical' to 'autobiographical' memory. The distinction here is between memory as directly experienced on the one hand, and memory as indirectly stimulated and mediated through wider agencies and ideologies (Coser, 1992).

For Nora, however, transition is much closer to rupture. His view of memory is as something organic and holistic, while history, by contrast, is a reconstruction and representation of the past. Dealing with our current condition, Nora sees 'real' memory in near-prelapsarian terms, whilst 'modern' memory consists of an array of historical traces. The key signifier of this shift to modern memory is the appearance of the trace. And with the onset of the dominance of history, he argues, real environments of memory, out of which memory arises spontaneously, can no longer be sustained. Consequently memory needs to be artificially created, fixed and represented in the form of *lieux de mémoire*; material, symbolic and functional at one and the same time. The nature of these sites of memory is evocatively captured by Nora when he writes of: 'moments of history torn away from the movement of history, then returned; no longer quite life, not yet death, like shells on the shore when the sea of living memory has receded' (Nora, 1989, 7–9, 12).

Halbwachs's conceptualization of the means by which memory is activated in the present relies on the notion of 'landmarks'. These, he argues, engender recollections and act as prompts for action in the present. Collective memory 'does not preserve the past but reconstructs it with the aid of the material traces, rites, texts and traditions left behind' (Halbwach, 1992, 175). This reconstruction, however, is always undertaken in the context of the present. Memorials and the process of memorialization are important means to this, as commemorative activity is one of the principal ways by which 'historical memory' is stimulated.

For Savage (1994), Nora's understanding of the relationship between history and memory is flawed. In the context of a persuasive exploration of the politics of collective memory, particularly as they relate to the creation of monuments to the American Civil War, Savage takes issue with Nora's dichotomy of internal and external memory. Nora argues that the modern tendency to deposit memory in external 'traces' has displaced and depleted 'real' internal memory. Savage raises two objections to this view. Firstly, he argues that all shared memory regardless of its location requires 'mediating devices'. This objection is predicated on the belief that Nora's real memory is shared. Savage's second objection is that the modern reliance on memory traces 'does not mean that more ephemeral and less easily documented means of remembering have been abandoned' (1994, 146). Rather, he believes that these two memory networks are not mutually exclusive in the way Nora seems to suggest. In his study of the memorialization of the American Civil War, Savage recognizes the continued existence of internalized forms of memory making such as veterans' reunions, alongside the creation of external traces in the form of monuments, reliquaries and archives. From this he argues that it may well be better to view these networks as 'mutually reinforcing', albeit in the context of the emergence of a 'hierarchy of memory activities' associated with the rise of the nation state and nationalist demands for tangible memory artefacts (Savage, 1994, 146). It remains to be seen whether monument making in the late-twentieth-century Scottish Highlands supports this view of the rise of the nation state bringing with it the privileging of tangible memory artefacts.

The Lewis Cairns

Much of the recent literature on Highland heritage has followed the well established and pessimistic view of the deployment of the past in the service of the present. For the Highlands, Harper and Vance (1999) emphasize the adoption of a romanticized and idealized Scottish past for contemporary commercial advantage. They focus on the role of the past in the present-day tourist industry and the evolution of a 'Theme Park nation'. Finally, while they admit that heritage can empower, they argue that only rarely does it empower women (Harper and Vance, 1999; Gibson, 1996). One of the aims of this section, however, is to demonstrate that the Lewis cairns, despite the dissonance that inevitably accompanies heritage making, represent a significantly more optimistic form (Map 2.1).

The five cairns that are the subject of this chapter ostensibly share a common intent, namely, the memorialization of the nineteenth and twentieth-century agitation for land, although in many other aspects they are significantly and symbiotically different. The origins of each of the cairns can be traced to the 1980s when a group calling itself *Cuimhneanchain nan Gaisgeach* (Commemorating Our Land Heroes) was set up. It is critically important to appreciate, however, that only three memorials were built by *Cuimhneanchain nan Gaisgeach.* Independent groups specific to the area in which the memorials were to be located built the other two. This split is made especially manifest in the built form of the memorials. Those built by *Cuimhneanchain nan Gaisgeach* are explicitly abstract and representational forms of public art (Figures 2.1, 2.2 and 2.3). By way of sharp contrast, the other two are wholly vernacular. It is argued here that these differences represent the inevitable contestation and conflict that emerges in any form of identity and heritage making and especially around the question 'whose heritage?' The remainder of this chapter focuses on the processes surrounding this politics of representation.

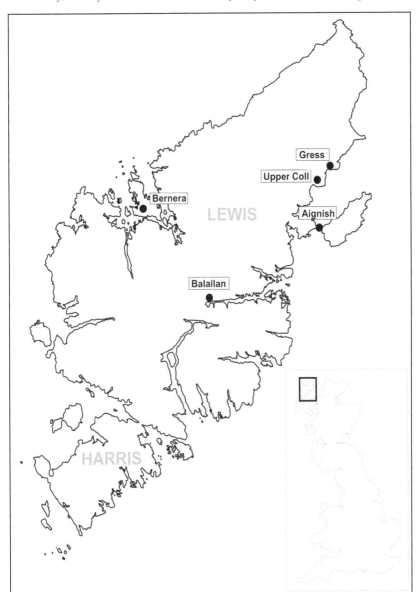

Map 2.1   Location of Memorial Sites on Lewis

As we have already noted, the memorials to both the Clearances and the Highlanders response to them have received much attention in the academic literature. Basu (1997) focuses on clearance commemoration as part of an explicit attempt to examine and interrogate landscape narratives. Drawing on the work of Christopher Tilley, he argues that 'the concrete details of locales in the landscape' serve in part to give 'stories' about places their mythic value and historical relevance. In terms of the clearance 'story', he recognizes the existence of a 'dominant narrative' that draws history, myth and memory into one and which is sutured into tangible memorials to these events. In turn, these memorials act as mnemonics for this constitutive narrative. Memorials for Basu, then, can function as a means to 'performative remembering', that, in the Highland context at least, aid the making and maintaining of an origin myth for Highland identity. Indeed, Basu goes further than this and argues that stories told 'in place do not only create a sense of identity, but also shape that identity in practice' (Basu, 1997).

Withers (1989) too, in his analysis of the commemoration of protest in response to socio-economic change, places much weight on the constitutive role of memorialization in memory making and local identity. He, however, sees this as more complex and less linear than Basu appears to suggest, and claims to be unable to find a single collective memory at work here. Nevertheless, Withers is clear that it is popular and local rather than public dominant or elite memory that is written into these memorials. What these commemorations represent for Withers, and in the context of a wider set of commemorations than those under consideration in this chapter, is the reassertion of local memory in place. He does, however, caution against viewing this working out of local memory as uncomplicated and uncontested. In the remainder of this section we will demonstrate precisely how complicated and contested the reassertion of local memory through the Lewis memorial cairns actually was and continues to be. Our intention here is, first, to document the commemorative process by discussing how the cairns came to be built in the manner and form they eventually took. Secondly, we intend to discuss the politics of representation surrounding these memorials, and, in particular, the way in which conflict, contestation and dissonance are bound up with these explicit attempts to make heritage landscapes.

In terms of the commemorative process, it has become apparent from interviews with members of the various organizing committees that while the final form of the monuments was the result of a collaborative endeavour, credit for the conception of the project and for ensuring that it became reality, was entirely due to the vision and motivation of one man, Angus Macleod. For Angus, inspiration came as early as 1986 when he was invited to Tiree to the opening of a memorial to the Reverend Macallum who had been influential in the fledgling land movement on Tiree. Although Macallum had subsequently been sent to Lewis, where he exerted a great deal of influence, it was the people of Tiree that chose to commemorate him and not those on Lewis. For Angus, the suggestion that Lewis was neglectful of its need to recognize influential events and people, was one that needed to be addressed and was further exacerbated by his awareness of memorial cairns to land disturbances on Skye. 'I said to one or two of my friends: "we are a shamed people, we're forgetting our history. Other people are looking after their history and shouldn't we not? There are at least four episodes here in Lewis which are well worth commemorating." So everyone was agreed and up and formed a committee and sailed off from then.'[3]

---

3   Interview *Jim Crawford*, Lewis, 8 April 1998; Interview *Angus Macleod*, Lewis, 6 April 1998.

Figure 2.1   Memorial to the Pairc Deer Raiders at Balallan
Source: Iain Robertson

Each subsequent act of memorialization was organized and directed by a committee, constituted for that event alone, with the intention of engendering a sense of ownership for each cairn in the area of the island to which it belonged. This ultimately ended in at least partial failure, however, an indication of the power and presence of dissonant heritage in even the most localized of landscapes.

The individual events that were chosen for commemoration were the riots at Bernera and Aignish, the Pairc Deer Raid and the meeting between Lord Leverhulme and the crofters from Back, Coll and Gress, an event which was itself consequent upon the long series of land raids in the area in the 1920s. Each of these events were of great significance in the campaign for land on Lewis and attracted considerable publicity and public interest (Buchanan, 1996). The events at Pairc and particularly at Bernera were of national significance in Scotland, and for Angus Macleod the events virtually chose themselves.

Each organizing committee was large and somewhat fluid, but included representatives of local government, the media and the Gaelic arts from the local areas where the cairns were to be located. The intention was that the memorials be erected sequentially, with a different chairman for each phase drawn from the district in which the cairn was to be erected. Angus was the first chair and the first cairn was built at Balallan. It was designed to commemorate the Pairc Deer raiders and was located in Lochs, Angus Macleod's home district (Figure 2.1).

Notwithstanding the clear intent behind the project, moving it towards construction proved difficult. This was mainly due to a lack of funding. The key to

Figure 2.2   Memorial to the Aignish Rioters
Source: Iain Robertson

carrying the project forward and successfully funding it proved to be the decision to adopt an overt public art agenda. Much of the impetus for this came from the director of the local arts trust who was a founder member of the project and from the decision to co-opt Roddie Murray, the Director of the *An Lantair* Art Gallery, onto the committee in 1992. At first unsure of his role, Murray found a voice arguing for a form of memorial that was something more than the vernacular. He argued that 'cairns should not be just cairns [...] in the sense that a cairn is a pile of stones [...] it was imperative that each memorial should reflect the relationship between the site, the events that had occurred at the site and the issues that had generated those events'.[4] For Murray, public art was the only form capable of achieving this. It would have been possible to have used different media to make the same point but this would not have reflected the strong connection between place and event. Indeed, it is possible to argue that, in terms the events these memorials celebrate, place, in effect, created event. Moreover, any alternative form of memorialization would not have created such lasting reminders within the landscape, visual markers that people would pass by every day. Finally, a vernacular pile of stones would emphatically not achieve the overtly representational mode deemed necessary, as they would not speak to a constituency wider than the purely local.

Some members of the committee, however, argued that the memorials ought to operate on a local level, and this fundamental difference in opinion generated a great

---

4   Interview *Roddie Murray*, Lewis, 7 April 1998.

Figure 2.3   Memorial to the Coll and Gress Land Raiders at Gress
Source: Iain Robertson

deal of tension within the committee. It also created tension between the committee and two of the recipient areas and ultimately caused a fracturing of the entire project. In the first instance, this tension was manifest in the argument that the public arts route would inevitably result in 'some modern artist [being] parachuted in' from outside and that the result would be 'something that they didn't understand and was irrelevant. They all had prefixed ideas about art, about what modern art in particular was, and none of them liked it.' Despite these misgivings the committee agreed to talk to the eventual artist, Will MacLean, and were somewhat reassured by the fact that he had Gaelic antecedents and was steeped in the history of land issues on the island.[5]

What eventually ensured the transformation from vernacular cairn to public art memorial was Angus Macleod's conversion to the latter approach and the fact that the Gulbenkian Foundation agreed to part fund the project, providing that the arts approach was adopted.[6] The result was three, acclaimed, public art memorials (Figures 2.1, 2.2 and 2.3), the final form of which was the product of a reflexive, interactive process of negotiation between the artist, a structural engineer, whose role it was to render the artists designs capable of being built, and a stone mason. Importantly, however, this was not a linear process from conception to final product. Rather, each of the triumvirate had a creative input. This is illustrated in particular in the choice of building stone. The preferred material was Lewisian Gneiss, indigenous

5   Interview *Roddie Murray*, Lewis, 7 April 1998.
6   Interview *Roddie Murray*, Lewis, 7 April 1998.

to the island and easily obtainable, and yet where this was sourced became of deep significance to the project as a whole.[7]

These memorials, as with the 'Power of Place' project in Los Angeles, are explicit attempts to make hidden histories visible (Hayden, 1995; Lacy, 1993). Moreover, they at first appear to fit with what Gillis (1994) has identified as an 'avant-garde counter monument movement' based in the belief that traditional forms of memorial making induce 'forgetting rather than remembering', through a passive engagement with the past and through a failure to induce questioning. The counter monument movement strives to do the opposite through the use of 'radical design' (Gillis, 1994, 16–17). On Lewis there is no doubt that active engagement and questioning became one of the ambitions of *Cuimhneanchain nan Gaisgeach*, even if it was not one from the outset. For Angus, the cairns were designed to make people, both local and visitors, question their purpose. '[The cairns] wouldn't be anything if they were heaps of stone [...] no they had to be something that would last, that would be there and that people would ask "well, what's that for?"' And whilst these memorials were explicitly designed to provoke such questioning, the designs were also firmly rooted in a deep understanding of the place in which they were built and were therefore also 'designed by history'.[8] This dualism perhaps brings into question the placing of these monuments into the counter-monument movement category and certainly points toward the fact that one of the principle objectives behind the cairns was to materialize memory in a fixed and highly condensed space.

It is doubtful whether the second major aspect of the memorialization process, the opening ceremony, fits as well with the second main aspect of the counter monument movement's approach, that of an explicit attempt to deritualize and dematerialize remembering. At both Balallan and Aignish the official opening was preceded by a march to the commemoration site. Furthermore, at Balallan this march closely followed the route taken by the participants in the events being commemorated. Here, local children were at the forefront of the event, whilst at Aignish the centrepiece was the performance of a play re-enacting the events commemorated. In both cases, therefore, ritualized acts were deployed in a deliberate attempt to capture history, relocate it into the twentieth century, turn it into a commemorative act and thereby create heritage. These are certainly not attempts to deritualize and dematerialize remembering and they may well have fixed forgetting.

While this may be the case it does not detract from the fact that these memorials were a deliberate attempt to create 'symbolic spaces endowed with significance because of their association with ideologically charged events that have been selected, polished and memorialised' (Osborne, 1996, 26). Despite this, the notion of *lieux de mémoire* perhaps fails to capture the counter-hegemonic spirit of the project. More persuasive here may well be the use made by Keith and Pile (1993) of Jameson's notion of cognitive mapping. They view the uneven development of space – produced by the logic of capital itself – as something of an opportunity.

7   Interview *Crawford*, 8 April 1998; Interview, *John Norgrove*, 8 April 1998.
8   Interview *Macleod*, 6 April 1998.

Mapping of these spaces by oppositional cultures permits their utilization as 'sites of resistance' (Keith and Pile, 1993, 3–4). This is a point underlined by Sherman (1994) who, drawing on the work of Raymond Williams, argues that local artefacts of commemoration can enjoy considerable autonomy from these wider structures, processes and ideologies (Sherman, 1994, 186–87). This, however, may not be as readily apparent in the Lewis cairns. The fact that *Cuimhneanchain nan Gaisgeach* was granted and then accepted a Civic Trust award for design must raise some doubt over the autonomous nature of the project. Nevertheless, there is no doubt that whilst the cairns were not explicitly constructed as sites of resistance they do celebrate a culture of resistance and of counter hegemony. Moreover, the ideas, centring upon attitudes to land that underlay protest and counter hegemonic social formations and which may be said to both derive from and constitute the 'taskscape', remain strongly evident in the contemporary crofting community (Ingold, 1993; Edensor, 2002; Robertson and Richards, 2003). This is clearly evident in the comments of local crofter, Marissa Macdonald:

> I think in a community like this your sense of identity, your sense of culture is the only thing that keeps you here. I walk down the croft which has been in our family since the sixteenth-century and even as a young girl I used to walk down the croft and think if only everybody, I mean it was home, it was always home to me its still home to me […] It's the same ground I'm tilling today to plant potatoes and its the same ground they tilled, same footsteps, same places. [pause] It gives you a very strong sense of yourself and who you are and where you come from. there's nothing worse than having that vacuum in your life […] [Land] is the only thing that keeps us clinging on by our fingernails to these rocks, these windy rocks.[9]

What these beliefs constitute, although not universally shared, is what Graburn (2001) identifies as the 'symbolic estate' of the crofting community. This notion encompasses two key features: firstly, that of cultural inheritance – the acquisition throughout one's life of an estate, material and non material, believed to be one's own by right, by descent, by expectation. For Highland crofters land has certainly formed part of their 'estate'. Secondly, Graburn argues, this idea – the sharing of a symbolic estate – is a way of understanding how the personal (inheritance) becomes the communal, and feeds into identity (Graburn, 2001, 71). This seems to suggest that these memorials are rooted in and central to an uncomplicated and uncontested local, crofting identity. This, however, is not the case, as both heritage and identity carry with them the potential for conflict and dissonance. In this instance these memorials both deny the inheritance of those who chose not to take part in protest and ignore the fact that Highland social protest was not wholly generated by consensus within the crofting community but by conflict also (Robertson, 2002). It is apparent, therefore, that conflict, fracture and disjuncture is as much characteristic of the making of these memorials as is cooperation. Most importantly, conflict and dissonance appears to have centred on questions of the role, significance and meaning of memorial making to local identity.

---

9   Interview *Marissa Macdonald*, Lewis, 6 April 1998.

Figure 2.4   Memorial to the Upper Coll Land Raiders
Source: Iain Robertson

As we have already noted, the intention was to build four memorials, each of which was to have an opening ceremony. And this ceremony was perceived as critically important in the passing on of the memorial to the local community. *Cuimhneanchain nan Gaisgeach* was unable to wholly fulfil either of these intentions. In particular, the committee encountered problems with both the Back and Bernera monuments. At Back, problems initially centred on the design of the monument and a considerably revised version was eventually built (Figure 2.3). The land disturbances celebrated by this memorial involved people from at least three separate, but contiguous, townships. Events of protest throughout the area were coordinated to occur at virtually the same time and were certainly treated as one united effort by agencies of government with whom protestors corresponded as a single group. Notwithstanding this, members of the individual townships felt that these events should have been memorialized individually. At one time there had been a representative from Upper Coll on *Cuimhneanchain nan Gaisgeach* but this was no longer the case at the time the details of the Back monument were finalized. People from Upper Coll consequently felt disenfranchized by this, and further alienated from *Cuimhneanchain nan Gaisgeach* by the public art form of the proposed monument and by the position of the site chosen. The site was selected for its local significance, as it was the location of an important meeting between the crofters of the area and the then proprietor of the island, Lord Leverhulme. It was a meeting that did not break the impasse between the two sides, a fact reflected in the design of the monument, and which strengthened the crofters resolve to continue with their acts of protest. Despite the importance of this location, committee members

Figure 2.5   Memorial to the Bernera Riot
Source: Iain Robertson

from Upper Coll felt alienated by the fact that it was closer to Back and Gress than their township. Their response was to erect an alternative monument in Upper Coll (Figure 2.4). Such was the strength of this opposition and the dissipating energy of the alternative monument, that any momentum behind an opening ceremony for the public art memorial was lost.[10]

At Bernera, the intended fourth site, as with Upper Coll, there was tension over both the nature of the memorial to be built and *Cuimhneanchain nan Gaisgeach*'s role in any memorialization. There was a strong body of opinion in the area that the public art aspect of the planned *Cuimhneanchain nan Gaisgeach* memorial was inappropriate for the purpose of the memorialization. Far more fitting would have been a vernacular cairn (Roddy Murray's 'traditional heritage of the Highlands') as this is the 'traditional way of celebrating big events' and 'more in keeping with the nature of the Lewis people'.[11] As with all the other memorials, people from Bernera had been observing *Cuimhneanchain nan Gaisgeach's* proposal for at least two years and had become increasingly uncomfortable with what seemed to them to be the committee's ever-more elaborate plans. The result was that at a series of public meetings the memorial was rejected. A Bernera-based group was formed; they raised funds within the area and erected a traditional cairn and dedication plaque (Figure 2.5). This was not done out of parochialism. The idea of a joint Lewis-wide

---

10 Interview *Macleod*, 6 April 1998; Interview *Macdonald*, 6 April 1998; Interview *Kenny McIver*, Lewis, 24 January 2005.

11 Interview *Murray*, 7 April, 1998.

celebration was understood and appreciated but what wasn't appreciated and was actively disliked was 'a highfalutin' artwork designed by someone from outwith the Lewis community'. This is the same reaction members of the *Cuimhneanchain nan Gaisgeach* committee had when the public art approach was first suggested. For Roddy Murray, the cause of Bernera's continued rejection of the *Cuimhneanchain nan Gaisgeach* memorial is to be found in their 'own very territorial reasons'. For members of the group charged with creating their own memorial, attempts by Angus Macleod and other members of the *Cuimhneanchain nan Gaisgeach* committee to persuade them to change their minds only pushed the locality in the opposite direction. As Noreen McIver says 'it became quite political'.[12]

For Roddie Murray from Coll, *Cuimhneanchain nan Gaisgeach* were rejected by the Bernera people because they felt imposed upon by those who did not 'belong' to Bernera and that it should be Bernera people who were in control of the process of memorialization. This is further exemplified by the fact that the Coll and Gress memorial was only accepted and built because he, Murray, was from the district and was able to convince the district that this was something of worth. Angus Macleod, although still Chair of the committee, felt it best not to attend the meeting at which Roddie Murray persuaded the local community to accept the committee's design, as Macleod had had an earlier experience of local parochialism:

> He says [to me] 'there is no point someone outwith the area going down there' as he (Angus) had been told at a meeting shortly after the end of World War Two that he had no right to advise them as he did not come from the area. What underlay this disjuncture, then, was traditional rivalry between districts on the island and a sense of (very) local identity.[13]

For Roddie Murray, Balallan 'was fine because Balallan was Angus's own home patch and ... John M. Macleod was also on the committee from there'. The monument at Upper Coll was built to act as a reminder 'to the people in the village themselves [...] a memorial for future generations of the village'. There is a connection to be made between island-wide events but each individual village has to have its 'own story'.[14] Tension involving the Bernera and Back/Upper Coll cairns, then, exemplifies attempts to maintain distinctive and very local identity. In both cases, moreover, the failure to accord with the *Cuimhneanchain nan Gaisgeach* project, meant that the two cairns separately became, perhaps unintentionally and unconsciously, inward looking.

At one level, then, this conflict demonstrates the difference and tensions that emerge when monuments are deployed as either inward-looking or outward-looking mnemonics. The Bernera riot had significance beyond both Bernera and Lewis and nationally for the land protest movement as a whole. An outward-looking public art memorial would have signified this. The consequence for the memorializations as a whole would have been to strengthen their claim to wider Gaelic identity making.

---

12 Interview *Noreen McIver*, Bernera, 4 February 2005.

13 Interview *Murray*, 7 April 1998.

14 Interview *Murray*, 7 April 1998; Interview *McIver*, 24 January 2005.

But, by creating an inward-looking monument the cairn sets up a dissonance between the event and its memorialization and celebrates only the local community and its identity. Yet, the process of memorialization revealed here demonstrates also the multi-layered nature of heritage and identity making. Angus Macleod also felt strongly that as the memorials 'are about our history' then they had to be 'from and by and for the locality'. And while they may also help a 'stranger' understand the local community, they were not built for the stranger but for the locality. His locality is not that of the individual area on Lewis but is island-wide. It was important also that the memorials were built of Lewisian gneiss as this was significantly more meaningful 'than getting a big slab with correct writing on it from away. It's local built by local people with local stone'. As John Norgrove articulates 'by the very nature of what they are made of they're a reflection of the place itself'.[15] And it is taken to another level by the stonemason for whom getting the right, historically significant, stone became a near obsession. This is particularly the case for the Balallan memorial, for which some of the most important stones came form the Bernera district. This, for Jim Crawford, became doubly significant in the context of the rejection of *Cuimhneanchain nan Gaisgeach* by the Bernera people. 'If the memorial could not go to Bernera at least Bernera was going to the memorial.'[16]

For Roddie Murray the memorials do their memory work on a number of different levels. It is important that they were 'cairn-like' in their design but equally important that the art aspect took them to 'another dimension'. They worked internationally as attractions and access points into the historical sub-cultures of the Hebrides and Lewis and Harris. The land disturbances were 'fulcrums of history'. These monuments celebrate fulcrums. Both history and heritage whilst crucially being about the making of local place are, at one and the same time, also about wider place making and contextualization. The process of memory making that is the Lewis memorials is one of the 'beautiful history of the island marred, because we just can't agree among ourselves'.[17]

Conclusion

Literally, figuratively and in the sense utilized by Halbwachs (1992), the Lewis memorial cairns as built by *Cuimhneanchain nan Gaisgeach* and others are deliberate acts of landmark making. The aim is to transform an otherwise unremarkable landscape into a psychic terrain; symbolic spaces that fix, or attempt to fix, collective remembering and act as prompts for a shared identity.

Landscapes represent power and control. Those with the power write their own landscapes in their own image, although it would be facile not to accept that contestation and conflict can also be written into a landscape. What we must also caution against is accepting these categories of dominance, contestation and conflict

---

15 Interview *Norgrove*, 8 April 1998
16 Interview *Macleod*, 6 April 1998; Interview *Crawford*, , 8 April 1998
17 Interview *Macleod*, 6 April 1998; *Murray*, 7 April 1998

as monolithic and unilinear. The meanings written into landscape are both plural and multilayered. There can be no doubt, however, that the crofting landscape is both the material manifestation and representative of the introduction of capitalist relations into the north west Highlands and islands. However, and echoing the fluidity which we must now accept surrounds the making and maintaining of meanings in a landscape, the crofting landscape also grounds those values and meanings that were to generate acts of social protest, attempts to recover lost land and, ultimately, the memorials that celebrate and seek to perpetuate the memory of these acts.

It is entirely appropriate, then, that these memorials celebrate a heritage, with its connotations of fluidity and multiplicity of meaning. Nevertheless, each of the Lewis memorializations, both those undertaken by *Cuimhneanchain nan Gaisgeach* and those not, are explicit attempts to create heritage from below, they are also explicit attempts at memorialization. They suggest a more fixed and self-contained form of heritage making. These memorials are attempts to fix and authentically record what is in the past and offer that past to future generations of insiders. They are, in short, attempts to create distinct landscapes of belonging. As such, these landscapes cannot be unproblematic. Where identity is made primarily with reference to the Other, the corollary of heritage is dissonance. Contestation is intrinsic to any attempt to write heritage into the landscape in the form of memorialization. Moreover, both heritage and identity are susceptible to contestation from within as much as they are from without. It would seem that the contestations around memorial making on the island of Lewis arise from traditional local rivalries and from a fear that the strong voices on *Cuimhneanchain nan Gaisgeach* were attempting to disinherit areas from which they did not originate. On Lewis there is a strong sense of identity based on where you come from. Neither the committee nor the memorials could and can overcome the mutual suspicion and sense of contestation generated by these identities.

Dissonance, identity and popular memory, as they are worked out in the landscape, are most often explored in the context of a national heritage. And yet we are compelled to agree with Withers (1989) that it is a matter of local rather than national identity being contested here. What Withers does not allow for is quite how local this contestation can get. Undoubtedly these memorials mediate the landscape and comprise a sense of identity at a very local scale. Nevertheless, this attempt to build heritage as commemoration has proved incapable of being able to transgress the boundaries created by these self-same identities. Dissonance is intrinsic to heritage landscapes even in this very localized context.

References

Ashworth, G.J. and Larkham, P.J. (eds) (1994), *Building a New Heritage*, Routledge, London.

Basu, P. (1997) 'Narratives In a Landscape', http://www.btinternet.com/~paulbasu/narratives/nl-frame.html.

Buchanan, J. (1996), *The Lewis Land Struggle*, Acair, Stornoway.

Coser, L.A. (1992), 'Introduction', in Halbwachs, M. (ed.), *On Collective Memory*, University of Chicago Press, Chicago, pp. 1–40.

Dicks, B. (2000), *Heritage, Place and Community*, University of Wales Press, Cardiff.

Dodgshon, R.A. (1998), *From Chiefs to Landlords*, Edinburgh University Press, Edinburgh.

Edensor, T. (2002), *National Identity, Popular Culture and Everyday Life*, Berg, London.

Gibson, R. (1996), *Highland Clearances Trail*, Highland Heritage Books, Evanton, Ross-shire.

Gillis, J.R. (1994), *Commemorations: The Politics of National Identity*, Princeton University Press, Princeton.

Gold, J. and Gold, M. (1995), *Imagining Scotland*, Scholar Press, Edinburgh.

Graburn, N.H.H. (2001), 'Learning to Consume', in Alsayyad, N. (ed.), *Consuming Tradition, Manufacturing Heritage*, Routledge, London, pp. 68–89.

Graham, B., Ashworth, G.J. and Tunbridge, J.E. (2000), *A Geography of Heritage*, Arnold, London.

Gray, M. (1957), *The Highland Economy, 1750–1850*, Oliver and Boyd, Edinburgh.

Halbwachs, M. (1992), *On Collective Memory*, University of Chicago Press, Chicago.

Harper, M. and Vance, M.E. (eds) (1999), *Myth, Migration and the Making of Memory*, John Donald, Edinburgh.

Hayden, D. (1995), *The Power of Place*, MIT Press, Cambridge, MA.

Hobsbawm, E. (1998), *On History*, Abacus, London.

Ingold, T. (1993), 'The Temporality of Landscape', *World Archaeology*, 25, pp. 152–74.

Keith, M. and Pile, S. (eds) (1993), *Place and the Politics of Identity*, Routledge, London.

Lacy, S. (1993), 'Mapping the Terrain: the New Public Art', *Public Art Review*, Fall/Winter, pp. 26–33.

Longley, E. (1991), 'The Rising, the Somme and Irish Memory', in Ni Dhonnchadha, M. and Dorgan, T. (eds) *Revising the Rising*, Field Day, Derry, pp. 29–49.

Lowenthal, D. (1985), *The Past is a Foreign Country*, Cambridge University Press, Cambridge.

Macinnes, A.I. (1996), *Clanship, Commerce and the House of Stuart, 1603–1788*, Tuckwell, Edinburgh.

McCrone, D., Morris, A. and Kiely, R. (1995), *Scotland – The Brand*, Polygon, Edinburgh.

Nora, P. (1989), 'Between Memory and History: *Les Lieux de Mémoire*', *Representations*, 26, 1, pp. 7–25.

Osborne, B. (1996), 'Figuring Space, Marking Time: Contested Identities in Canada', *International Journal of Heritage Studies*, 2, 1/2, pp. 23–40.

Robertson, I.J.M. (2002), '"Their Families had Gone Back in Time Hundreds of Years at the Same Place": Attitudes to Land and Landscape in the Scottish Highlands after 1914', in Harvey, D.C., Jones, R., McInroy, N. and Milligan, C. (eds) *Celtic Geographies*, Routledge, London, pp. 37–52.

Robertson, I. and Richards, P. (2003), 'Introduction', in Robertson, I. and Richards, P. (eds), *Studying Cultural Landscapes*, Arnold, London, pp. 1–12.

Samuel, R. (1994), *Theatres of Memory*, Verso, London.

Savage, K. (1994), 'The Politics of Memory: Black Emancipation and the Civil War Movement', in Gillis, J.R. (ed.), *Commemorations: The Politics of National Identity*, Princeton University Press, Princeton, pp. 127–49.

Sherman, D.J. (1994), 'Art, Commerce and the Production of Memory in France after World War One', in Gillis, J.R. (ed.), *Commemorations: The Politics of National Identity*, Princeton University Press, Princeton, pp. 186–212.

Tunbridge, J.E. and Ashworth, G.J. (1996), *Dissonant Heritage*, Wiley, London.

Withers, C.W.J. (1988), *Gaelic Scotland*, Routledge, London.

Withers, C.W.J. (1989), 'Place, Memory, Monument: Memorialising the Past in Contemporary Highland Scotland', *Ecumene*, 3, 3, pp. 325–44.

Withers, C.W.J. (1992), 'The Historical Creation of the Scottish Highlands', in Donnachie, I. and Whatley, C. (eds) *The Manufacture of Scottish History*, Polygon, Edinburgh, pp. 143–156.

Chapter 3

The Battle for Annie Moore: Sculpting an Irish American Identity at Ellis Island National Monument

Joanne Maddern

The Biography of an (Extra) Ordinary Immigrant

Annie Moore (1877–1923), a fifteen-year-old emigrant who left Ireland for the US in 1892, has become something of an unlikely iconic figure in American immigrant culture.[1] As the first immigrant to be processed at Ellis Island, New York, a facility that officially opened on 1 January 1892 to police incoming waves of migrants, she was admitted into the new world amid a 'festive din of bells and steamer whistles' (Kelly and Morton, 2004). After questioning by Charles M. Hendley, Secretary of the Treasurers Office, the well rehearsed narrative goes, she was 'set free' in America with a one-off gift of a ten dollar gold coin and the promise of a new life in a land of 'freedom and opportunity'.

Annie's biography and identity has since been heavily mobilized by the Irish and American heritage industries to symbolize the building of a new nation through the hard work, self-sacrifice, bravery and noble suffering of individual immigrants. Annie Moore's life history whilst fascinating, is rather unremarkable. It may even be described as *tragic*. In this way her story could not be more different to the celebrated life stories of notable 'public' immigrants like Lee Iacocca, the symbolic head of the Statue of Liberty – Ellis Island Foundation (SOL-EIF), the 'private not-for-profit' foundation entrusted with the task of generating public money with which the lauded multi-million dollar restorations of the Statue of Liberty and Ellis Island were achieved.

Juxtaposed with Annie Moore, Lee Iacocca is the archetypal migrant success story. A charismatic businessman and son of Italian parents, he saved the floundering Chrysler Corporation and fits snugly into that very particular category of famous

1   Thank you to John Walsh, Chairman of the Irish American Cultural Institute and various members of the National Park Service for granting interviews. I am appreciative of the NPS for allowing me to peruse the archives in New York, Washington and Boston. Thank you to Susan Kelly and Stephen Morton for passing on some of their work, pre-publication. This chapter was improved by comments received by Paul Bevan, Luke Desforges, Tim Cresswell and Sarah Cant. The ESRC has funded the larger body of research of which this was a part.

American immigrants who can be called upon to signify the core values of American social life such as ingenuity and prosperity and rugged individualism. Lee Iacocca is a modern day nation builder, who endured particular kinds of suffering and adversity to become an upwardly mobile upholder of the virtues of free enterprise, competition and ingenuity.

Annie's story, by contrast, goes a little differently. In 1892, aged just fifteen, she left County Cork and her native land of Ireland along with her two younger brothers, aged just seven and eleven.[2] She rode the *SS Nevada* for ten days across the Atlantic Ocean and disembarked along the gangplank on New Year's Day alongside 700 migrants from three ships moored in New York Harbour. Since Annie and her brothers were steerage passengers, they were required to undergo a battery of tests at the brand new processing station on Ellis Island.

By successfully negotiating the inspection procedures in the new landing bureau's great hall at Ellis Island, Annie and her brothers were eventually reunited with her waiting parents, Matt and Mary Moore, and her oldest brother, Tom, all of whom had emigrated two years before her in search of a better life. While in the US, Annie's parents had had two more children, Patrick and Elizabeth. By seeking a better future there, Annie and her family had followed in the footsteps of hundreds of thousands of Irish emigrants who had fled the 'great hunger' of the mid-nineteenth century.[3] Irish migrants (along with German migrants) had dominated the immigrant flow to New York from the mid to late-1800s (Foner, 2000). In leaving Ireland for a new life in America they were also blazing a trail for the continuing flow of Irish migrants that passed through Ellis Island in their thousands between 1892 and 1924. For many Irish emigrants, migration became part of a continuing struggle for national liberation. Irish migrants who arrived in large numbers in the mid-1880s for instance, were often deeply involved in the Irish nationalist cause of the early decades of the twentieth century. Foner (2000) has argued that New York migrants continue to be tapped by home-country politicians and political parties as a source of support, just as earlier in the twentieth century Irish nationalist politicians made pilgrimages to New York.

The Moore family lived in New York for a while, where a wealthy family employed Annie as a housemaid. Eventually they travelled west to Indiana where Annie met Patrick O'Connell, a descendent of Daniel O'Connell, a well-known Irish Statesman and patriot from County Clare. Annie married Pat O'Connell in Texas in 1898 aged twenty-one. The entire family settled in Waco, central Texas, to farm. Soon, Annie discovered her lungs were weak, so the O'Connell family moved to west Texas, where the climate was deemed more suitable for her weakening constitution. Annie and Patrick soon began a family, although only five out of the

---

2   There is some confusion concerning the names of the brothers travelling with her. Some newspaper sources document their names as Anthony and Philip, whilst others name them Joe and George.

3   It is thought for example, that between ten thousand and thirty thousand Irishmen died whilst digging the Pontchartrain Canals through the swamps of New Orleans.

eight children that Annie gave birth to survived. Mary Catherine, the eldest of the surviving children, is often depicted with her mother on commemorative plates sold at the Ellis Island gift shop.

Like billions of women around the world, Annie looked after her children without pay, public recognition or conventional accolade in the private and invisible sanctum of the home. She was widowed at an early age as a result of a flu epidemic that prematurely killed her husband. The bereavement occurred shortly after they had relocated again, this time to Clovis, New Mexico on doctor's recommendations. Her last child, Anna, was born the same year. Annie had been left to carry out the labour-intensive work of bringing up five children independently. Fortuitously, Patrick O'Connell had been a shrewd trader, and just before his death had acquired a portion of land in Clovis, including a hotel and restaurant profitably situated near the then thriving railroad. Annie and her older children managed the property, turning it into a substantial business after her husband's death.

In 1923, Annie decided to make a fateful trip to Fort Worth, Texas to visit her ill younger brother, Pat, who remained there with Annie's other brothers and sisters. Travelling between Dallas and Fort Worth, tragedy struck. Annie was hit and killed by an 'inner-urban' train, one of the first rapid-transit trains between the two cities. After her premature death aged just forty-six, her five orphaned children were sent back to Texas to be cared for by various aunts and uncles from the O'Connell family. Each child went on to receive a college education.

Although the journey of Annie and her brothers from Ireland to America was covered by the *New York Times* and New York's *The World*, her relatively short life was not particularly marked by exceptionality, lasting fame or fortune. Yet her name, appearance and identity have posthumously been heavily mobilized by the tourist industry as a potent symbol of Irish-American heritage. Among other things, it is now possible to buy several biographies charting Annie's life, a number of children's stories by Eithne Loughrey, fictionalizing Annie's place in American history, replicas of the *SS Nevada* ship manifests on which Annie is listed as a passenger and a variety of quintessentially Celtic Annie Moore dolls dressed in period clothing that sit alongside a wide variety of Irish-American memorabilia in the Ellis Island gift shop (Loughrey, 1999; 2000; 2001) (Figure 3.1).

One can also listen to songs about Annie Moore performed by the 'Three Irish Tenors', on their album titled 'Ellis Island', and drink a pint of authentic Irish Guinness in innumerable Annie Moore pubs scattered across North America from New York to Florida. Further, an 'Annie Moore Award' is presented annually at the Irish-American Cultural Institute's St. Patrick's Day Ball in Morristown, New Jersey to an individual who has been seen to have made 'a significant contribution to the Irish or Irish-American community and legacy'.

It might be said, then, that the once largely unknown Annie Moore has posthumously become something of a cult heroine in American immigrant culture. The Annie-Moore 'brand' has become a ubiquitous signifier of Irish-American identity embedded in numerous variations within the American cultural landscape. But significantly, it is only long after her death that Annie's ghost(s) have been conjured up and made to

**Figure 3.1   A Selection of Irish-American Memorabilia Displayed in a
                Cabinet in the Ellis Island Gift Shop**
Source: Joanne Maddern

'speak' by the heritage industry. Such posthumous fame is inherently problematic for
in her subaltern silence, it is difficult to hear precisely what she is trying to say. As
Kelly and Morton (2004, 121) suggest '[t]he figure of Annie Moore disappear[s] in a
series of dead ends, gaps and ellipses within an archive that represents an Irish woman
as an exemplary US citizen'. The remainder of this chapter presents a case study of
the 'identity work' carried out at Ellis Island through the excavation of Annie's life-
story and the erection of a sculpture in her name.

A number of researchers have commented recently on the ways in which museums
have struggled to represent the transnational identity transformations taking place
within global cultures. Macdonald for instance has identified 'arguments by social
theorists… [which suggest that] many former identities are in the process of radical
transformation, of fragmentation and disembedding' (Macdonald, 2003, 2–3).
She considers the implications of these radical transformations for museums, and
suggests that:

> if the nation-state and the kind of 'public' with which it was associated are on the brink of
> obsolescence, then what future is there for museums? Are museums perhaps too intimately
> linked up with material- and place-rooted, homogeneous and bounded, conceptions of
> identity to be able to address some of the emerging identity dilemmas of the 'second
> modern age' or 'late modernity'? […] museums, precisely because they have been so
> implicated in identity work and because of their more particular articulations within the
> kind of [nationalist] identities that are argued to be under threat, are significant sites in

Figure 3.2   A Life-size Bronze Sculpture of Annie Moore, Displayed in the Main Building on Ellis Island

Source: Joanne Maddern

> which to examine some of the claims of identity transformation […] We might expect that museums as institutions would become redundant […] Alternatively, we might expect to see transformations within museums as they attempt to address and express 'new identities'. (Macdonald 2003, 6)

The Ellis Island Immigration Museum can be used as a prime case study to explore how complex identities which exceed simple national classifications are 'managed' within a national museum. Whilst successive waves of transnational migration stretching back more than two centuries have shown that the identity transformations discussed by Macdonald are in no way an altogether recent or unique occurrence, museums as we will see, continue to struggle with geographically interconnected pasts that are deemed to exceed national borders and boundaries (Gilroy, 1993; Hall, 1990; Bhabha, 1990).

Elsewhere a variety of authors have provided a set of differently positioned readings of Ellis Island Immigration Museum, its exhibitions and displays through the lenses of heteronormativity (Rand, 2003); memory, power and nation building (Kirshenblatt-Gimblett, 1998; Wallace, 1996); historic preservation (Johnson, 1984); immigrant representations within the US (Bodnar, 1995; Desforges and Maddern, 2004); visitor expectations (Frisch, 1990); diaspora tourism (Maddern, 2004); place, pilgrimage and Jewish identity (Perec, 1995, 1999) and American social history (Smith, 1992).

Figure 3.3   A Section of the Annual St. Patrick's Day Parade, New York
Source: Joanne Maddern

Here, I draw specifically on a series of excerpts from letters and memoranda covering the period 1991–1993 and located in archives at National Park Service locations in New York, Washington and Boston. In particular, I want to analyze a particular battle that took place between National Park Service heritage professionals, political actors and Irish-American ethnic organizations.[4] The battle was concerned with the placement of a life-size bronze statue of Annie Moore in the refurbished main building at Ellis Island (Figure 3.2). This research examines the micro-geography of the struggles and battles that surrounded the presentation of a geographically interconnected history of one ordinary Irish-American woman at a well-known national heritage site.

Conjuring the Spectres of Annie Moore

There is a strong tradition of Irish migration to America since at least the eighteenth century (Coogan, 2000). A vibrant Irish-American diaspora exists in pockets across the US, as attested to by the animated annual St Patrick's Day Parade that takes place

---

4   From here on the National Park Service will be referred to as the NPS and the Irish American Cultural Institute will be referred to as the IACI.

every year in New York, a tradition that dates as far back as 1766 (Figure 3.3). The mid–nineteenth century influx of immigration through Ellis Island was dominated by those of Irish origins, creating a strong Irish-American collective identity. Drawing on media representations of 'the Irish world-wide', Gray has recently written of the ways in which, 'emigration occupies a contradictory position in Irish cultural memory as something to be forgotten and remembered depending on socio-political context' (Gray, 2003, 157). She argues that 'like many other emigrant nations in the 1990s [...] the Republic of Ireland was reclaiming its diaspora as a means of refiguring the national as global'. It was in such a context that the IACI decided to donate a bronze life size statue of Irish Annie Moore in honour of the one century anniversary of her crossing the Atlantic (Gray, 2003, 157).

The IACI coordinated an international effort to have a commemorative statue of Annie Moore sculpted and erected at the recently reopened Ellis Island, and at her Irish departure point, Cobh, in County Cork. The Cobh sculpture was to be part of the 'Queenstown project', a one million pound recreation of the story of Irish emigration from 1750, orchestrated by the Cobh Heritage Trust Ltd who promoted it as the 'perfect foil to the Ellis Island Centre' (Hogan, 1992, 7). In County Cork the project aimed to 'create again for visitors the precise conditions which the emigrants would have found when they arrived at the old railway station in the town from all over Ireland to board ship for what was very often a gruelling sea journey' (Hogan, 1992, 7). Cobh's history as a world-class seaport would also be documented at the Centre, reviving images of the old town. With the Queenstown project receiving around £7,000,000 from the European Union and Ellis Island undergoing a $156 million dollar restoration using donations from corporations, foundations and individual citizens, these two symbolic sites were set to once again reinvigorate their historic links with migration between the 'old world' and the 'new.' Plans were made for a 'coordinated ceremony with ... President Robinson ... asked to do the honours and ... a live link on the day [of the unveiling] with an exchange of greetings'.[5]

The campaign to erect the two memorials facing each other some three thousand miles apart on either side of the Atlantic Ocean was ultimately successful. In 1993, the incumbent Irish President Mary Robinson unveiled both of the life-size pieces of art, designed and sculpted by the renowned Irish artist Jeanne Rynhart, perhaps better known for her Molly Malone sculpture in Dublin. In the time between the idea for the statues and their final unveiling however, a significant but little known battle raged over whether Annie Moore's statue would be 'in place or out of place' (Cresswell, 1996) at Ellis Island national monument. This was a battle that incorporated an ever-widening circle of individuals and political, social and ethnic organizations, and a battle that resulted in a significant shift in the form that the statue would eventually take.

The battle centred on the following concerns: what would the aesthetics of Annie Moore's immortalized feminine form communicate to Irish-Americans and non Irish-Americans alike? Would her enshrined presence be appropriate at a NPS

---

5   Matt McNulty, Deputy Director General, Bord Fáilte (Irish Tourist Board) in a letter to Mr. John Walsh, IACI, 29 January 1992.

museum and monument? 'Statues are powerful catalysts of spectralisation', writes Hetherington, in that, they 'stand as a metaphor for the ghost in every object', part of the European tradition of erecting statues to 'honour or bear witness to the... heroes of the city's past for their great deeds' (Hetherington, 2001, 30). Yet Annie Moore was no 'great hero', at least not in the traditional sense of the word. As such, if the statue was included in the cultural landscape, it would provide a new historical opposition to the hundreds of commemorative monuments of 'powerful' political or military men to be found in National Park Service sites across the US.

Neither would a statue of Annie Moore be an 'empty vessel' upon which the flamboyant desires and meanings of those casting their detached gaze over her could impose, as with the omnipotent and iconic Statue of Liberty standing adjacent to Ellis Island in the New York Harbour. Her statue, if it was to be accepted by the museum, would be uniquely different from the preponderance of feminine statues symbolizing abstract, idealized qualities such as freedom, liberty and justice. While Annie had had an independent life and identity of her own as an ordinary teenager, how would her modest feminine stature sit alongside the material remnants of the old and elitist histories whose legacies remained in the national cultural landscape of the NPS? Though a revised thematic framework at the National Park Service had strengthened its commitment to the 'new social history' and placed great emphasis on 'ordinary people and everyday life' over and above 'great men and events', the prospect of the erection of a life-size statue of Annie Moore, an 'ordinary' woman, still had the power to cause considerable controversy (NPS, 2001).

Her life it seemed, did not fit with standard migration histories. It was not a straightforward tale of upward mobility. It contained tragedy in equal measure with triumph over adversity. Her identity as a wife, homemaker and mother of five was one built around the needs of the family and unpaid physical and emotional work rather than one that was publicly centred on free enterprise and rugged individualism. Most significantly (and somewhat controversially) her statue seemed to be able to invoke multiple connections that stretched outside the territory of the nation through her transnational ties with the 'old world'.

The campaign started positively. On 27 December 1991, John Walsh from the Irish-American Cultural Institute in New York penned a letter to Mr. James Ridenour, director of the National Park Service in Washington DC to introduce himself, the institute and the Annie Moore Project:

> You have received correspondence from Ambassador Moore and former Governor Kean about the strong interest in placing a life-size bronze statue of Annie Moore on Ellis Island to commemorate her place in history as the first person to be processed through Ellis Island on January 1, 1892. The project is gaining momentum and enjoys the support of the Irish Government. The noted Irish sculptor, Jeanne Rhynhart has agreed to cast the statue, which will show Annie Moore and her two younger brothers on a gangplank. Although arrangements are not finalized, we envision the finished sculpture to measure approximately 10ft high by 12ft wide by 6ft deep. We believe that the sculpture will be

both a visual and symbolic enhancement to Ellis Island. And of course, the project will be completely self supporting.[6]

A barrage of letters was duly received by the NPS throughout 1991 and 1992 supporting the project, echoing the optimistic tone of John Walsh's initial letter. For instance, Denis Malin, Consulate General of Ireland, argued that Annie Moore and her brothers represented 'the vanguard of that vast diaspora of [Irish] immigrants' that passed through Ellis Island.[7] Hugh L Carey, Executive Vice President of the Office of Environmental Policy wrote to Mr. Lujan, Secretary of the Interior asking that the NPS should 'look favourably on the request to place the Annie Moore statue on Ellis Island', because 'for millions of visitors from throughout America and the world she will symbolize the dreams and hopes that brought each new generation of immigrants through that historic gateway'.[8]

Former New Jersey Governor Thomas H. Kean likewise wrote to Mr. Lujan assuring him that 'whilst the Irish carry pride for her, Annie Moore's significance is not in being Ireland's first immigrant to America, but the very first immigrant from *any* country to pass through Ellis Island... [so that] her statue could stand as a source of pride to all sons and daughters of immigrants in America'.[9] In a similar letter later that year, he espoused these sentiments again, suggesting that the statue could stand as a 'poignant and powerful tribute to every immigrant who has knocked on the "golden door" in New York Harbour'.[10]

Jim Florio, Governor of the State of New Jersey composed a letter too, suggesting that the NPS should accept the offer of the statue immediately. He began, 'I am writing to you regarding the erection of a statue of Annie Moore on Ellis Island. In telling the Ellis Island story, it seems most appropriate to start from the beginning. Annie Moore is the beginning.'[11] Advocates of the statue made their opinions felt through a rapidly increasing stock of supporting letters and memoranda about the intrinsic links between Annie Moore's biography and the 'American Dream', in which the great wave of migration to the US played a central role:

Annie Moore, the first person to be processed through Ellis Island, is part of the Island's historic character. The Irish American Cultural Society believes that she is a representation

---

6   John P. Walsh, Chairman and Chief Executive Officer, Irish American Cultural Institute, in a letter to Mr. James Ridenour, Director of the National Park Service, 27 December 1991.

7   Denis Malin, Deputy Consul General, Consulate General of Ireland, New York in a letter to John Walsh, 28 June 1991.

8   Hugh L Carey, Executive Vice President of the Office of Environmental Policy in a letter to Mr. Lujan, Secretary of the Interior, 5 March 1992.

9   Thomas H. Kean, in a letter to Manuel Lujan, Secretary of the Department of the Interior, 26 February 1992.

10   Thomas H. Kean, president of Drew University in a letter to Mr. James Ridanour, Director of the National Park Service, 21 June 1992.

11   Jim Florio, Governor of the State of New Jersey, in a letter to Marie Rust, Acting Regional Director, United States Department of the Interior, National Park Service, 25 February 1992.

of America's immigrant heritage. Like all the other immigrants that passed through the Island's facilities, Annie Moore came to this country looking for opportunities and a better life. The cultural society proposes a statue of Annie Moore to be placed on Ellis Island to remind us of our own immigrant roots, and that we too can share in the American dream.[12]

Support for placing an Annie Moore sculpture on Ellis Island was forthcoming from overseas too, from such quarters as the American Embassies in Dublin and Paris. From Dublin, Ambassador Richard Moore ruminated:

> I would like to think that Annie Moore may have been a relative, but even without that extra dimension I believe the memorial would symbolize the multicultural character of America in a most effective way.[13]

In an earlier letter, Moore had already laid out the positive effect he believed the acceptance of the statue would have in terms of solidifying particular collective memories of America's immigrant past:

> When I served in the White House in the 1970s I became well aware of the wide and important role of the National Park Service not only in providing opportunities for recreation, but as a custodian of the symbols of American history and culture. In that respect, I believe strongly that the proposed monument would be an important reminder of the contribution which Irish-Americans have made to the United States.[14]

From Paris meanwhile, Ambassador Walter Curley tried to assuage concerns that the sculpture would be prioritizing Irish identities over the many ethnic groups that passed through Ellis Island:

> The proposal is *not* singling out a particular country. It is singling out a particular individual: the *first*. A *personalized* memorial is much more representative and touching than a generic tribute covering 'The millions of immigrants who...'. As former Ambassador to Ireland, I can be accused of a certain bias. I stand guilty. But also I feel strongly about the basic merits of the Annie Moore Memorial in this Bicentennial Year for Ellis Island.[15]

Officials however were quick to refuse the offer of the statue because it did not fall within the park's established criteria found within the NPS 'Scope of Collection Statement'. NPS officials believed (despite all protestations to the contrary) that to accept it would single out a particular nationality to the exclusion of others:

---

12 Frank R. Lautenberg, United States Senate, Washington DC in a letter to James M. Ridenour, Director, National Park Service, 18 March 1992.

13 Richard A. Moore, Ambassador, Embassy of the United States of America, Dublin in a letter to Manuel Lujan, Jr, Secretary of the Interior, 24 April 1992.

14 Richard A. Moore, Ambassador, Embassy of the United States of America, Dublin, in a letter to James Ridenour, Director of the National Park Service, 15 November 1991.

15 Walter J. P. Curley, Ambassador, Embassy of the United States of America, Paris in a letter to Manuel Lujan, Jr, Secretary of the Interior, 2 April 1992.

We believe that it is doubtful that a statue of Annie Moore would be perceived as representing all other immigrants. Indeed, park visitors of other than Irish descent would likely view Annie Moore's statue as implying that their ancestors were some how of lesser importance. Inevitably there would be requests for statues to commemorate the immigrant experiences of Italians, Germans, Poles and other nationalities... Her story... is not inherently more important than the stories of the millions of other immigrants who braved adversity, danger, and the unknown to begin a new life in a new country.[16]

Furthermore, Marie Rust, NPS Regional Director wrote to supporters of the statue informing them that the NPS could not accept the statue since Annie Moore could not be considered of 'transcendental importance' to the interpretive history of the site. She wrote:

Placing a statue of Annie Moore at Ellis Island would be contrary to the National Park Service policy of not allowing commemorative statutory in national parks. Exceptions are made only when the association between the park and the individual is of transcendent importance. While Annie Moore was the first immigrant to pass through Ellis Island, her association with the site, we believe, in no way meets the criteria for the placing of a commemorative work at the park.[17]

John Walsh, Chairman of the Irish American Cultural Institute notes how the IACI, 'ran into difficulty with the National Park Service. They did not want to recognize any immigrant group over another immigrant group... It was a very difficult time and we had to get former governors, many senators, and many congressmen all to endorse it.'[18] He stepped up his campaign, undeterred by the initial refusal, explaining how he saw *Irishness* as something that could be easily incorporated into a national US site. This is an excerpt of a letter penned to Senator Lautenberg of New Jersey in a bid to drum up public backing for the statue:

...Annie Moore, a 15 year old from Co Cork, Ireland... and her family became quite successful in America. Isn't that what the American Dream is all about? Unfortunately, the Interior Department does not see it this way. It has rejected our proposal on the basis that it would be inappropriate to memorialize an Irish national, and to memorialize one nationality out of many... The fact of her Irishness is something in which we Irish Americans can take particular pride; as a symbol of the 100 million Americans of all races, creeds and nationalities who have descended from Ellis Island immigrants, Annie Moore is representative.[19]

---

16 Marie Rust, Acting Regional Director, NPS in a letter to Honorable Frank R. Lautenberg, United States Senate, Washington DC, 27 March 1992.

17 Marie Rust, Acting Regional Director, NPS in a letter to Honorable Frank R. Lautenberg, United States Senate, Washington DC, 27 March 1992.

18 Interview with John Walsh, Chairman of the Irish American Cultural Institute, March 2002.

19 John P. Walsh in a letter to Frank Lautenberg, 3 February 1992.

Yet, Marie Rust of the Park Service stood firm arguing that the 'placement of a statue commemorating an Irish national – even one as noteworthy as Annie Moore would establish the precedent of accepting commemorative sculpture in the future from other nationalities and ethnic groups, many of whom outnumbered the Irish for Ellis Immigration':[20]

> If we were to recognize Annie Moore as the first immigrant to pass through Ellis Island, would not the stories of the last, the youngest, and the oldest immigrants to pass through merit equal recognition? To take this a bit further, what about the immigrant who was 'housed' on the island for the longest period of time prior to being allowed into the country? Or, what about those immigrants who became well known Americans? [Or, those who were forced to return home?][21] Or those born on Ellis Island? Or those who died there… [or in the crossing]?[22] …Ellis Island is not about commemorating specific immigrants, it is about commemorating all immigrants.[23]

She concludes her rationale:

> [We do not] want to see the Monument cluttered with statues that place the focus on individuals rather than the many and varied people who passed through these portals. Our policy is intended to protect the intrinsic values and resources of the individual units of the National Park System, and to minimize contemporary intrusions.[24]

Finally, the political pressure applied through continuous letter writing overwhelmed the NPS. Yet, despite extensive archival research, how or why the statue eventually got there in spite of powerful opposition remains a mystery. The reason(s) behind this sudden change of decision by NPS officials, like the finer details of Annie Moore's life, remain a well-kept secret. In conjuring up the spectres of Annie Moore, the backstage-process of Ellis Island's production, and the absences contained within the NPS archive that records it, are also made visible.

The bronze Annie Moore statue was accepted into the museum's migrant landscape and unveiled by incumbent Irish president Mary Robinson in a dedication ceremony in 1993. The sculpture depicts Annie Moore with one hand on her hat (to hold it on her head while she lifts her head to take in the sights of New York Harbour), the other holding a small suitcase as a symbol of her immigrant journey. Though the statue took a slightly less flamboyant form than originally envisioned (Annie cuts a lone figure, her brothers and the gangplank do not make an appearance)

---

20 Marie Rust, in a letter to John Walsh, 18 January 1992.

21 This bracketed sentence appeared in the draft copy but not the final version of the letter.

22 This bracketed sentence appeared in the draft copy but not the final version of the letter.

23 Marie Rust, in a letter to Edward M. Kennedy, United States Senate, Washington DC, 6 May 1992.

24 Marie Rust, in a letter to Edward M. Kennedy, United States Senate, Washington DC, 6 May 1992.

it can be said that the letter writing and campaigning for the inclusion of the statue had been worth it. The IACI now describe the transatlantic Annie Moore statues as 'symbols of the permanent union between Ireland and the United States [and]... a symbol for immigrants of all nations that have contributed to the rich fabric of the United States'.[25] Annie Moore is now repeatedly invoked as a central protagonist in the human drama of transatlantic migration, a key figure in the globalization of Irish culture: 'Like the 12 million immigrants who came to Ellis Island after her', official descriptions proclaim, 'Annie Moore came to America bearing little more than her dreams. She stayed to help build a country enriched by diversity.'[26] Yet Annie Moore's placement within the museum continues to reflect her somewhat marginal and contested status. She does not occupy a central or prominent position on the island. Instead, the statue is discretely tucked away between two pillars inside the main building. It has no spotlights shining on it, and must compete for visitor attention with the large and arresting black and white photographs of migrants hanging on the wall directly behind it.

Conclusion: Material Cultures of Migration

The empirical material presented in this chapter has raised several interesting theoretical points: firstly, the ability of several different sets of social actors to affect the final display through an organized letter writing campaign is contrary to theories advanced by heritage critics who have too often suggested that museums are unproblematic representations of dominant elite interests and histories (Wright, 1985; Fowler, 1992). That said however, a counter argument could be made that Irish-American interest groups are a relatively power-enabled set of social actors in comparison to the power bases afforded by other immigrant groups such as, for instance, African-Americans or Asian-Americans (see Hoskins, 2004 on Asian Immigration history at Angel Island, nicknamed 'the Ellis Island of the West'). Gray (2003) notes for instance, how the (controversial) trope of 'empire' has been used to describe the 'new world status of *Irishness*'. She describes this ascendancy as an 'empire of the imagination' established through high levels of emigration, rather than an empire forged through domination. In this way the provocative use of the term empire is used unsettlingly to illustrate encounters between colonialism, postcolonialism and global capitalism. She writes:

> Music, dance, stories of famine and migration, but also practices of business and marketing [and we might add museum narratives and material cultures] are just some of the technologies of memory that produce the 'new world status' of 'the Irish'. Ireland and

---

25 Irish-American Cultural Institute. Annie Moore: The Annie Moore project, available at http://www.iaci-usa.org/anniemoore.html [accessed on 4 August 2003].

26 American Park Network. Statue of Liberty National Monument. History: A Closer Look. Webpage available at: http://www.americanparknetwork.com/parkinfo/sl/history/annie.html [accessed on 4 August 2003].

Irish culture are cast within 'new geographies of Imperialism'...that are structured by neo-liberal and free-market practices as well as new paths of cultural flows and intermixing. (Gray, 2003, 158)

Nash (2002) too, has examined the relative prominence of Irish-American genealogies within national histories in the US. Nevertheless, despite the prominent visibility of 'Irishness' within the United States, the ability of disparate groups of social actors to influence the material cultures of the museum points to the fact that museum landscapes are sometimes much more fragmented and disputed sites than often imagined.

Secondly and perhaps most interestingly is the question of 'transcendent importance', a term coined by the NPS to decide a historic individual's relevance to a larger historical narrative. 'Transcendent importance' signifies an almost metaphysical epistemology. For something to be transcendental, it must surpass the limits of ordinary knowledge and experience. It is in this supernatural space that history is open to considerable interpretation. To define an individual as transcendentally important to a national history is to bestow them with almost 'saint-like' qualities. This is something that the NPS has been historically happy to do with political and military leaders and other elites, but a new focus on 'social histories' within the Park Service problematizes such a definition. The bestowal of transcendent importance within a national narrative of belonging is further problematized when an individual has multiple affiliations and links to more than one country, as Annie Moore did.

Annie Moore's statue was considered by members of the National Park Service to be 'too Irish' to be easily assimilated into the museum space. Though during her lifetime Annie's migration to the US signified a desire to become an American, and her successful passage through Ellis Island meant that she had officially become a US citizen, in the eyes of the NPS officials, her enduring Irish-American links and affiliations precluded her statue from being immediately accepted in a national museum almost a century later.

The NPS desire to develop exhibits in a restoration that reflected a supposedly 'authentic and balanced telling of the Ellis Island story'[27] troubled the appropriateness of any aesthetic object in the museum that could have been seen to embody enduring connections and affiliations with particular 'old world' countries. Perhaps, all too fresh in the National Park Service collective psyche, was the historical centrality of the homelands in the nineteenth and twentieth century migrant imagination. Jacobson writes of the ways in which the first wave of migrants to the US avidly followed news of and remained actively involved in home-country politics. He argues:

Life in the diaspora, remained in many ways orientated to the politics of the old centre. Although the immigrant press was a force for Americanization, equally striking is the tenacity with which many of these journals positioned their readers within the envisioned nation and its worldwide diaspora.... In its front page devotion to Old World news, in its focus upon the ethnic enclave as the locus of US news, in its regular features on the groups'

---

27 Marie Rust in a letter to John Walsh, 24 January 1992.

history and literature, in its ethnocentric frame on American affairs, the immigrant journal located the reader in an ideological universe whose very centre was Poland, Ireland or Zion. (Jacobson, 1995, 2)

All too recent was the often uneasy process of Americanization through which the 'tired', the 'poor' and the 'huddled masses yearning to breathe free'[28] were sculpted into responsible American citizens (a process in which the NPS played a central role as custodian of the symbols of American history and culture). The NPS thus unconsciously saw the unassuming sculpture of Annie Moore as an impediment to the continuing project of Americanization in operation at cherished historic landmarks. These sets of geographically linked sites act as receptacles of collective memory through which national senses of social cohesion, belonging and relatedness are sustained and consolidated.

A rampant gendering of the past also permeates this narrative. Annie's nation-building contribution had involved bringing up and caring for a family of five and had been carried out largely in the private spaces of the home. Whilst a Park Service policy of 'not adding statues perpetuating the memory of specific persons in park service areas'[29] was cited as one of the many reasons for initially refusing the offer of the statue, there are clearly many examples of such statues in NPS sites across America. Statues of historically powerful political and military personnel adorn many National Park Service locations in the US. Whilst Annie's physical and 'emotional labour' was as valuable as any work carried out outside the home, to a NPS used to memorializing war heroes and political leaders and just starting to grapple with a new social history, her transition from the private space of the home when alive, to the very public spaces of a museum after her death, seemed unnatural and incongruent.

It was not an immigrant statue *per se* that the NPS objected to then, but what that statue may *represent* to Ellis Island visitors. The issue of representation, as we have seen, has been a matter of considerable consternation. The battle over the Annie Moore statue had lasted for over two years, and had involved an ever widening circle of local, national and international stake-holders. Each stake-holder had expressed different views concerning what they believed Annie's immortalized and aestheticized form would represent to museum visitors. For the NPS, it seemed that the nuances of Annie's everyday life (in the past) could not be smoothed out and dissolved into a narrative of a national past in which an American geographical imagination is given ontological priority.

Though the statue was eventually accepted by museum officials whose resolve was seemingly weakened by the tidal wave of letters supporting the Annie Moore project, the lengthy battle for its inclusion is symbolic of the complicated relationship that often exists between the subjectivities of well-developed diasporic identities and

---

28 This is from the sonnet 'The New Colossus', written by Emma Lazarus in 1883 and inscribed at the base of the Statue of Liberty. The phrasing belies the ambivalent way that European migrants were viewed at that time.

29 Michael Adlerstein, National Park Service, in a letter to John Walsh, 16 March 1992.

the official histories adopted by 'national' tourist sites. Annie's ghost unwittingly occupied an uneasy space lost somewhere between 'Irish' and 'American' identities nearly a century after Annie's death. The statue caused concern because conjuring up the spectres of Annie Moore risked an invocation of multiple connections and affiliations that stretched well outside the territoriality and temporality of the nation-state. As Kelly and Morton (2004) have explored, in 'calling up' Annie Moore, it is possible to invoke a number of biographical narratives which *contest* as much as reinforce the American dream thesis.

In returning to the questions posed at the start of this chapter, it seems fair to suggest that museum narratives *can* sometimes be too inextricably entangled in 'old' forms of identity to be able to express 'new' ones. But the substantive material presented here also complicates Macdonald's (2003) arguments about what types of histories and material cultures are constitutive of 'old identities' and what sorts comprise a 'new identity'. Annie Moore's statue can be seen both as representative of a transcultural set of knowledges and memories, and as embodying a staunchly Irish past. It could be said that in making an attempt to resist the inclusion of the statue, the NPS *were* in fact endeavoring to present new, hybrid forms of geographical identity and history though new innovative (non-individualistic) technologies of memorialization. On the other hand, the initial rejection of the statue can also be viewed as a refusal to incorporate the diasporic identities that Annie's statue clearly projects. The ability for objects, artifacts and sculptures to act in ways that are ambiguous and reflective of 'old' (nationalist) and 'new' (transcultural) identities simultaneously is underplayed in Macdonald's work, though she does suggest that 'trying to create historical accounts that eschew national or ethnic narratives as well as causal or progressive trajectories is a difficult task; and one that needs to be tackled through aesthetic strategies... as well as through content' (Macdonald, 2003, 10). This begs the question, can the types of identities and memories favoured by the 'new social history' ever be represented in a personalized fashion, or do they demand new technologies of representation?

In a period in which many theorists are arguing that the identities of the past are becoming increasingly irrelevant and new transcultural and transnational identity formations are being created, it makes sense for museums to once again critically examine the kinds of identities they are constructing for their visitors, not just through textual displays, but also through their material cultures of migration.

References

Bhabha, H. (ed.) (1990), *Nation and Narration*, Routledge, London.

Bodnar, J. (1995), 'Remembering the Immigrant Experience in American Culture', *Journal of American Ethnic History*, 15, 3, pp. 18–27.

Coogan, T. P. (2000), *Wherever Green is Worn: The Story of the Irish Diaspora*, Random House, London.

Cresswell, T. (1996), *In Place/Out of Place: Geography, Ideology, and Transgression*, University of Minnesota Press, Minneapolis.

Desforges, L. and J. Maddern (2004) 'Front Doors to Freedom, Portal to the Past: History at the Ellis Island Immigration Museum, New York', *Journal of Social and Cultural Geography*, 5, 3, pp. 437–458.

Foner, N. (2000), *From Ellis Island to JFK: New York's Two Great Waves of Immigration*, Yale University Press, New Haven.

Fowler, P. (1992), *The Past in Contemporary Society: Then, Now*, Routledge, London.

Frisch, M. and D. Pitcaithley (1990), 'Audience Expectations as Resource and Challenge: Ellis Island as a Case Study', in Frisch, M. (ed.) *A Shared Authority: Essays on the Craft and Meaning of Oral and Public History*, State University of New York Press, Albany, pp. 215–224.

Gilroy, P. (1993), *The Black Atlantic: Modernity and Double Consciousness*, Harvard University Press, Cambridge, Mass.

Gray, B. (2003), 'Global Modernities and the Gendered Epic of the "Irish Empire"', in Ahmed, S., Castaneda, C., Fortier, A. and Sheller, M. (eds) *Uprootings/ Regroundings: Questions of Home and Migration*, Berg, Oxford, pp. 157–178.

Hall, S. (1990), 'Cultural Identity and Diaspora', in Rutherford, J. (ed.), *Identity. Community, Culture and Difference*, Lawrence and Wishart, London, pp. 223–237.

Hetherington, K. (2001), 'Phantasmagoria/Phantasm Agora: Materialities, Spatialities, and Ghosts', in *Space and Culture*, 11/12, pp. 24–41.

Hogan, D. (1992) 'The Emigrant's Return', *The Irish Times*, Tuesday, 25 August.

Hoskins, G. (2004) 'A Place to Remember: Scaling the Walls of Angel Island Immigration Station', *Journal of Historical Geography*, 30, 4, pp. 685–700.

Jacobson, M. (1995), *Special Sorrows*, Harvard University Press, Cambridge, Mass.

Johnson, L. (1984), 'Ellis Island: Historic Preservation from the Supply Side', in *Radical History Review*, 28–30, pp. 157–168.

Kelly, S. and S. Morton (2004), 'Calling Up Annie Moore', *Journal of Public Culture*, 16, 1, pp. 119–130.

Kirshenblatt-Gimblett, B. (1998), 'Ellis Island', in *Destination Culture: Tourism, Museums, and Heritage*, University of California Press, Berkeley, pp. 177–187.

Loughrey, E. (1999), *Annie Moore: First in Line for America*, Mercier Press, Cork.

Loughrey, E. (2000), *Annie Moore: Golden-Dollar Girl*, Mercier Press, Cork.

Loughrey, E. (2001), *Annie Moore: New York City Girl*, Mercier Press, Cork.

Macdonald, S. J. (2003), 'Museums, National, Postnational and Transcultural Identities', *Museum and Society*, 1, 1, pp. 1–16.

Maddern, J. (2004), 'The Isle of Home is Always on Your Mind: Subjectivity and Space at Ellis Island Immigration Museum', in Coles, T. and Timothy, D. (eds), *Tourism, Diasporas and Space: Travels to Promised Lands*, Routledge, London, pp. 153–171.

Nash, C. (2002), 'Genealogical Identities', *Environment and Planning D: Society and Space*, 20, pp. 27–52.

NPS (2001) *History in the National Park Service, Themes and Concepts, Revised Thematic Framework*, US Department of the Interior, Washington, DC.

Perec, B. (1999), 'Ellis Island: Description of a Project', in Sturrock, J. (ed.) *Species of Spaces and Other Pieces*, Penguin, London, pp. 134–138.

Perec, B. and R. Bober (1995), *Ellis Island*, New Press, New York.

Rand, E (2003), 'Breeders on a Golf Ball: Normalizing Sex at Ellis Island', *Environment and Planning D: Society and Space*, 21, pp. 441–460.

Smith, J. (1992), 'Exhibition Review: Celebrating Immigration History at Ellis Island', *American Quarterly*, 44, pp. 82–100.

Wallace, M. (1996), *Mickey Mouse History and Other Essays on American Memory*, Temple University Press, Philadelphia.

Wright, P. (1985), *On Living in an Old Country*, Verso, London.

Chapter 4

Constructing Famine Memory: The Role of Monuments

John Crowley

Introduction

In the mid-1990s there was a general recovery of public memory with regard to one of the most tragic periods of Irish history, the Great Famine (1845–1852). The Famine was widely commemorated not only in Ireland but also amongst the Irish diaspora. The commemoration took many forms, which included television documentaries, radio broadcasts, museum exhibitions, monuments, walks, conferences, lectures and the creation of websites. Contemporary cultural geographers are interested not only in the sites/places/loci of memory but also in its contested nature. For example, societies can take very different meanings from the Famine and consequently the response to it can vary over time and place. Diasporic versions, for instance, can differ profoundly from the more nuanced versions to be found at present in Ireland. It is clear that the ways in which the Famine is remembered and commemorated is deeply implicated in issues of nationality and collective identity. This chapter will concentrate in particular on the politics of Famine memory by focusing on the role of monuments in the cultural landscapes of Ireland and the Irish disapora. Specific attention will be paid to monuments in Cork, Mayo, Dublin and Boston. To begin, though, I want to explore the fate of one monument of empire which provides an insight into the highly symbolic and contested nature of Famine memory in Ireland.

Famine Memory and the Legacy of Queen Victoria

In 1849 the principal architect of University College Cork, Sir Thomas Deane, presented the college with a statue of Queen Victoria. It occupied a central position in the University, on a plinth over the eastern gable of the Aula Maxima. Its chequered history from that date reflects the changing political climate in Ireland. With the emergence of an independent state in 1921 there was little time and space for monuments that reminded the people of their colonial past. The statue of Queen Victoria was removed in 1934 and buried in the grounds of the college. It was replaced by a statue of Saint Finbarr sculpted by the well-known Cork sculptor, Seamus Murphy. In many ways, this was a decision very much in keeping with the mood of Catholic triumphalism that prevailed in the newly established Irish Free State.

In 1995 University College Cork celebrated its 150th anniversary and a decision was taken by the governing body to re-erect the statue of Queen Victoria. Michael Mortell, then president of the university, proposed that it would become part of the Universitas exhibition which set out to document the history of UCC from its foundation to the mid-1990s. The decision provoked an angry response from a number of groups and individuals. A series of letters appeared in the local newspaper, the *Cork Examiner*, condemning the decision on the grounds that the Queen had presided over the most tragic period in Irish history. The thrust of the argument put forward by those opposed to the University's decision to exhibit the statue was captured in a series of letters submitted to both the local and national press. In the course of these letters, the Queen was variously described as 'a tyrant responsible for so much misery' and the statue was 'regarded as a symbol of British arrogance' (*Cork Examiner*, 30 December 1994). Those objecting to the re-erection of the monument, also argued that it was offensive to the memory of the IRA volunteers who were also buried in the grounds of the University (Murphy 1995).

The University, however, defended the decision on the grounds that the 150th anniversary of its founding was an appropriate time to re-examine its own past. The statue, it argued, would become part of an exhibition which would deal solely with the history of the institution. Virginia Teehan, the College Archivist, acknowledged the sensitivities of those who were strongly opposed to the idea:

> The concept of what was appropriate started to dominate all my waking and sleeping hours. The decision to display the statue was not an easy one and was not taken lightly. It was a difficult decision for the college officers and for the overall institution and when it was formally announced at a press launch in December most of my colleagues did not believe it. (Interview, Teehan, 1995)

As a piece of sculpture the statue is not of great artistic merit but its symbolism runs deep (Figure 4.1). Teehan sought to deflect attention by widening the discussion to include attitudes towards symbols on the island of Ireland in general in what was then a burgeoning climate of hope fostered by the Peace Process in Northern Ireland. Given that context, people could afford to look at monuments that were previously ignored or neglected in a new light. For example, it is only quite recently in the Republic of Ireland that attention has been focused on monuments commemorating Irishmen who died in World War One. Teehan further explained:

> The central issue here to my mind is not the statue itself but how we as a society look at symbols of the past. Monumental symbols like statues have a rhetoric. These rhetorical functions or meanings shift as society shifts and changes. This is not to say in any way, that the meanings or associations, which different generations have with an object, are forgotten or undermined. They become part of our larger collective memory. (Interview, Teehan, 1995)

Figure 4.1   Statue of Queen Victoria as Displayed in the Universitas Exhibition, UCC in 1995

Source: Virginia Teehan

Monuments like that of Queen Victoria provide an opportunity for the present generation to look at aspects of the past which have literally been buried or indeed obfuscated by time. In Irish folklore, Victoria, along with Sir Charles Trevelyan came to personify official British indifference towards Ireland during the Famine years. Even today it is difficult to change that perception. As Gibbons points out,

> the belief that the restoration of Queen Victoria's statue was an inoffensive gesture in the context of an historical arc spanning 1845-1995 could only make sense if the Great Famine in Ireland was a thing of the past, a phase of history that could now be safely confined to the communal Prozac of the heritage industry. But can the wounds inflicted by a social catastrophe be so easily cauterised? Would anyone suggest that the traumatic lessons of the Holocaust should not be as pertinent in a hundred years time as they are today. (Gibbons, 1996, 172)

Trying to escape the restrictive influence of such images and perceptions became central to the work of sculptor John Behan, whose Famine Ship project also raised many issues and fuelled debates about the contested nature of memory and commemoration.

John Behan's Famine Ship

> The Great Irish Famine of 1845–52 is embedded in the folk memory of all Irish people as
> an unresolved phenomenon – memories of history lessons at school are of unspeakable
> sadness and suffering to me personally. Like the generations before me, my peers and
> I have not faced up to the facts of the Famine. Perhaps the complexities of the Famine
> have not been properly laid out for us to understand in total. We have tended to highlight
> the incidents such as Queen Victoria's meagre contribution, something we can deal with
> rather than view in a detached way the reasons why the Famine occurred. (Behan, 1995)[1]

1995 marked the sesquicentenary of the Famine and the Irish government embarked
on a series of commemorative events and memorial projects. One such memorial
project was awarded to the sculptor John Behan. He was commissioned to create a
monument in memory of those who perished during the Famine. Behan, who works
in the medium of bronze, carried out a great deal of research into the history and
folklore of the period before ultimately deciding on the form of the monument. The
proposed site of the memorial in Murrisk at the foot of Croagh Patrick in County
Mayo had a profound influence on the artist. The mountain has been a focal point
for pilgrims for many generations and Murrisk is the village from which pilgrims
begin their ascent. Behan undertook his own pilgrimage, which he admitted was
more artistic than religious. From the summit of the mountain, Behan looked out on
a landscape and seascape that captivated him. Relic features in the landscape such
as the potato ridges and the empty shells of houses also recalled the impact of the
Famine years. His sculpture would later reflect the twin influence of land and sea.

   In a television documentary about the making of the memorial, Behan described
how he visited many Famine sites in order to assemble a series of images that he
could draw on later. His visit to the workhouse in Birr had a particular hold on
him. While walking through the various sections, he was struck by the crowded
conditions that must have prevailed in the workhouse during the Famine years. The
confined spaces of the women's sleeping quarters along with the timber-framed roof
inspired the image of the coffin ship. As Behan explained in the documentary, 'the
interior felt and looked like the interior of a coffin ship'. The other feature in the
landscape which had a particular resonance for him was the lazy bed:

> potato ridges built spade by spade, now lying under the grass inscribing the Famine into
> the very landscape. You learn to read the landscape, how it is marked by human habitation
> and how to incorporate that into your work.

As an artist Behan struggled in his efforts to come to terms with the scale of the
disaster. He explained:

> Death is impersonal when it strikes in such great numbers. We struggle to find a single
> human signature in the midst of such devastation – some simple way of comprehending

---

1   Behan described his artistic journey in a documentary from the RTÉ *True Lives* series,
entitled *Famine ship*, first broadcast in 1999.

Figure 4.2　John Behan's Famine Ship, County Mayo
Source: Yvonne Whelan

a large catastrophe. An artist's journey is a bit like that, trying to find the human scale in the suffering of a whole people. The image of the ship has deep ramifications throughout human history and human culture. As a symbol it is very much associated with the journey of life and the journey of death.

The landscape around Clew Bay in Co. Mayo would remain uppermost in the artist's mind as he refined his image of the famine ship. As was highlighted in the *True Lives* documentary, Behan 'regarded it as a place haunted by final journeys and absences. These overlapping themes would become a central concern in his work. The texture and colouring of the Famine Ship in the end would mirror the landscape of ridges, rivulets and bare rock'.

For famine victims, the ships that would take them away from Ireland proffered hope, but also embodied the fear of the unknown. The stark image of the skeletal figures, which make up the rigging of Behan's Famine Ship, reveals in no uncertain way the sad fate of the many who tried to leave the horror of the Famine behind them (Figure 4.2). Each art object, be it a painting or sculpture presents its own hermeneutical challenge. They either provoke a response or not. It is only recently while reading an account of the Irish Famine in Simon Schama's *A History of Britain* that the power of Behan's sculpture became more obvious to me:

In Connemara on the Atlantic shore, it seems to have been the father's task to take their dead babies to the edge of the ocean to the ancient limbo spaces of water, land and sea;

and dig little graves, marked by a rough stone cut from the cliffs. Circles of 30 or 40 of the wind-scoured, lichen-flecked stones, their jagged grey edges pointing this way and that, stand by the roaring surf, the saddest little mausoleum in all of Irish history. (Schama, 2001, 304)

The Famine Ship memorial has been described as one of the most powerful commemorative symbols created during the 150th anniversary (Longley, 2001). As O'Toole explains, 'no art can do justice to the suffering of famine victims and any attempt to do so runs the risk of pathos. But what is too monstrous to be adequately described can be evoked' (O'Toole, 1997, 10). Behan's Famine Ship is a memorial that defies stereotypes. It both illustrates history and humanizes it. It takes an imaginative approach, which evokes the horror of the Famine in a very powerful and meaningful way.

Commemorating the Famine in Dublin

The recent spate of commemorative events relating to the famine, especially since the mid-1990s, disguises the fact that the traumatic events of the mid-nineteenth century had long been over-looked by officialdom. In Ireland's capital city, Dublin, the famine was not officially commemorated until 1967 when Edward Delaney's Famine memorial was unveiled. It was one of a series of works sculpted by Delaney, which also included monuments of Theobald Wolfe Tone and Thomas Davis. The Famine memorial (Figure 4.3) was positioned in a quiet corner of St Stephens Green, separated by a granite wall from the Wolfe Tone monument on the other side that looked out onto the busy Merrion Row. Delaney's sculpture afforded the memory of the Famine a physical presence in the capital city, albeit one which was initially eclipsed by the imposing figures of Davis and Tone. Given that 1966 marked the fiftieth anniversary of the 1916 Rising, it was perhaps not altogether surprising that the Famine should have taken such a back seat in the commemoration stakes. This had also been the case in 1945, the year that marked the one hundredth anniversary of the Famine as well as the centenary of Thomas Davis's death. While the Davis commemoration took centre stage the Famine was relegated to the margins. The bleak socio-economic conditions that prevailed in the new State help explain the rather muted official response. As Daly explains:

> the strong and unqualified belief in Irish nationalism during the 1940s is immediately obvious from the uncritical attitudes paid to figures such as Davis or Parnell. Although the Famine is often seen as giving a boost to the cause of Irish independence, the fact that the centenary was ignored suggests that the event conveyed some less comforting messages, such as the failure of the independent Irish state to reverse a century of emigration and population decline, or its difficulty in feeding and heating its citizens adequately during 1947. (Daly, 1995, 17)

In this case, however, the official response merely reflected a wider public apathy. Indeed the neglect of the Famine in this period lends weight to Declan Kiberd's

Figure 4.3   Edward Delaney's Famine Memorial, St. Stephen's Green, Dublin
Source: John Crowley

contention 'that far from being worshippers of the past, what Irish people really worship is their own power over it, including the power to bury it at a time of their own choosing' (Kiberd, 2000, 651). It goes without saying that along with the power to bury the past comes the power to resurrect it. In this respect the State's and indeed the wider public's willingness to embrace the commemoration of the Famine in 1995 stands in marked contrast to 1945. Gray and Oliver explain the recent surge of interest in terms of the pursuit of cultural stability in a rapidly changing society:

> While the symbols of a largely Catholic, inward-looking and autarkic nationalism may appear increasingly problematic in Ireland's secularising and internationalising society, there remains a demand for some historical continuity, a collective identity, rooted in a distinctive Irish past; and the Famine appears to many to offer a focus that is at once catastrophic, local, diasporic and relevant to the modern world. (Gray and Oliver, 2001, 12)

Recently commissioned artistic works on the theme of the Irish Famine must be seen within this rapidly changing context, and it is to one of these that I wish to now turn.

Rowan Gillespie's Famine Memorial was unveiled in 1997 in front of Dublin's Custom House. The emaciated figures which line the quayside along the banks of the river Liffey, recall in no uncertain terms the tragedy of the Famine and they are meaningful in a way which is quite difficult to articulate (Figure 4.4). Plaques surround the figures, inscribed with the names of the memorial's sponsors, the

Figure 4.4   Famine Memorial, Custom House Quay, Dublin
Source: John Crowley

majority of them drawn from Ireland's rich and powerful business elite. It had originally been the artist's intention to include the names of individuals and families who had perished during the Famine years. This proposal was later rejected, however, by the organizing committee in favour of its own fund-raising venture and sponsorship initiative. For a minimum contribution of £5000, companies from all over the world were invited to pay tribute to the Great Irish Famine 'by having their company name cast in bronze on one of the many flagstones along the docks of Dublin city, a place where many left during the Famine era' (Humphreys, 1998, 4). One organization which took great exception to this initiative on the grounds that it demeaned the memory of the Famine and it's victims was Action for Ireland group, commonly known as AfrI. The problem with the memorial, they argued, lay not with the sculpted figures themselves but with the way in which the memory of the Famine was being commodified. AfrI argued that:

> there is something inappropriate in the idea of selling advertising space to companies around the base of the monument. The use of the memory of the Famine dead as a means of boosting corporate and individual egos is to say the least gross and insensitive.[2]

---

2   O'Toole makes the argument that the lack of sensitivity was lamentable. However, what was even more regrettable in his opinion was the fact that the insult didn't register with the organizers, see O'Toole, 1998, p. 14.

Figure 4.5   Detail of Famine Memorial, Custom House Quay, Dublin
Source: Niamh Moore

But it is not just within Ireland that commemoration of the Famine has become a contested topic. One of the greatest and lasting impacts of the Famine has been the emergence of an extensive Irish diaspora, none more vocal than in the Irish American heartland of Boston, Massachusetts.

Commemorating Famine Overseas: The Boston Memorial

Over three-quarters of the one million people who fled Ireland during the Famine years made their way to the United States. It naturally follows that amongst Irish Americans the Great Famine represents a substantial and significant communal memory. The institutionalization of that memory has become a dominant theme in recent years. Here, I want to explore one particular commemorative project, the unveiling of a monument in Boston. The Boston memorial was designed by Robert Shure, and comprises two life-size sculptures, one depicting the desperate plight of a family leaving Ireland, the other of a family arriving in Boston determined to succeed in their new surroundings. The origins of the project can be traced back to the early 1990s when then Mayor of the city, Raymond L. Flynn put forward a plan for a memorial to be located close to Faneuil Hall marketplace. After some preliminary work the project lost impetus in the wake of Flynn's appointment as US ambassador to the Vatican. In 1996 a local real estate developer, Thomas J. Flatley, resurrected the idea and put in place a committee that would drive the project forward. This was made up of a consortium of business leaders, scholars, writers, clergy and representatives from various Irish-American organizations.[3] Located at the corner of Washington and Scholl streets, near Downtown Crossing, the memorial was to be situated near the Old South Meeting House and was to be part of the city's Freedom Trail. This trail attracts in the region of 16 million visitors annually. Given the centrality of the location, the memorial park represented one of the most ambitious attempts by Irish Americans to commemorate the Famine and to inscribe its memory on the city's streetscape.

The monument was unveiled on 28 June 1998 when an estimated 7,000 people attended the dedication. The memorial narrative is predicated on sources such as John Mitchel, Lady Jane Wilde and William Carleton and the underlying message is one of the triumph of the Irish in the face of adversity, an oppressed people had not only escaped the hunger but also the callous treatment meted out to them by the British government and found freedoms in America that had been denied to them in their homeland. America was the haven, which allowed the emigrants the opportunity not only to live but also to prosper. It is a narrative, which speaks more of success than failure, the American dream writ large. The *Boston Globe* editorial on 9 March 1998 commented:

> Beyond its particularly Irish dimension, the memorial marks the beginning of the waves of 19th and 20th century immigration that have made Boston the variegated place it is today. Thousands more would come: Italians, Jews, Greeks, Lithuanians, Chinese, Haitians, Dominicans, blacks from the American South and other ethnic groups, all seeking a refuge from poverty and oppression. The triumph of the Irish is a parable of America.

The Famine, however devastating its consequences, marked the beginnings of the inexorable rise of the Catholic Irish in America. It could be argued that the emigrant experience is not really dealt with adequately in Shure's monument. To present that experience in such a positive light misrepresents the often tough and uncompromising conditions which the emigrants had to endure both on their journey to the New World and on their arrival. The monument seems preoccupied with the constant striving forward rather than marking with dignity the memory of the Famine victims and the harsh conditions faced by many Irish emigrants in their new surroundings. The striving forward has become an integral part of the Irish-American's self image. In many respects the monument is more a tribute to the achievements of the Catholic Irish in the United States than saying anything profound about the Famine itself. According to cultural critic Fintan O'Toole, the monument paints a rather simplistic picture of what was a complex tragedy. He makes the argument that the lack of sensitivity was lamentable. However, what was even more regrettable in his opinion was the fact that the insult didn't register with the organizers (O'Toole, 1998). In that context the triumphal note it sounds at the end seems to diminish the real suffering of those who perished during the period. O'Toole was trenchant in his criticism of the Boston memorial, arguing that: 'the actual monument is a dreadful piece of

---

3   For the background to the monument, see http://www.boston.com/famine/history.stm.

kitsch. Beautifully crafted kitsch certainly, expensive kitsch – it cost $1 million – but kitsch nonetheless. It shows not an ability to face our past, but a complete inability to imagine it. As a memorial to the dead, it offers pious clichés and dead conventions. As an effort to confront a national trauma it shows a depressing immaturity' (*Irish Times*, 3 July 1998).

Irish America is not a monolith but is made up of many communities. What the Famine really means to the people who make up such communities is beyond the scope of this chapter. However, if as Paul Ricoeur suggests, cultures have founding events then the Famine represents such an event for a significant number of Irish Americans (see Kearney, 1992). For the many thousands who fled Ireland in the 'Hungry Forties', and in particular their descendants, the Famine marks the beginning of their story. It remains the touchstone of their identity as a people. The question of which version of the Famine becomes dominant in particular places and times is very much linked to the question of identity. The Famine as exile is very much the sustaining myth of Irish America. Myth in this context is not used in a pejorative sense but in the wider context of explaining 'how meaning is constructed out of a catastrophic event' (Howells, 1999, 4). The fact that a culture nurtures certain images of the Famine cannot be so lightly dismissed. It could well be argued that O'Toole's criticism of the Boston monument misses the point in not analyzing more deeply the spaces and contexts in which the memorial is located. The dissonance that now characterizes debates about the meaning of the Famine in Ireland and the United States reflects in the end the very different needs of complex and diverse societies.

Conclusion

The focus of this chapter has been on the role of monuments in shaping the memory of the Famine. According to Johnson, 'monuments are an important source for unravelling the geographies of political and cultural identity especially as they relate to conceptions of national identity' (Johnson, 1995, 52). The very different responses to the display of the statue of Queen Victoria as part of University College Cork's Universitas exhibition in 1995 reflected not only shifting and conflicting attitudes to the past but also the complex nature of Irish society and identity in the 1990s. It is clear that as societies change the meaning of monuments can also change. The construction of public memorials to the Famine reveals a great deal not only about the aesthetics of memory but also significantly its politics. The poverty that prevailed in Dublin in the 1940s, for instance, can explain the rather muted official response to the 100th anniversary of the Famine. In a period of heightened nationalism – 1945 also marked the centenary of Thomas Davis's death – the Famine took a backseat in the official commemoration stakes. A State commissioned Famine memorial did not materialize until 1967 and even then it was part of a trilogy of works by the sculptor Edward Delaney in which memorials to Wolfe Tone and Thomas Davis occupied the more open civic spaces. The Famine memorial's comparative invisibility in a sense reflected the State's unease with the Famine and its memorialization.

This ambivalence stands in marked contrast to both the Irish government's and indeed the wider public's response to the Famine more recently. A rapidly changing political, socio-economic and cultural context provided a more favourable climate for the State and indeed the public to engage with the memory of the Famine. For example, the old certainties of Irish identity (Nationalism and Catholicism) have begun to unravel and in such a climate there is no alternative but to construct in the words of Gillis (1994, 20) 'new memories and new identities better suited to the complexities of a post-national era'. The renewal of interest in the Great Famine and indeed other historical events such as the Great War must be seen in the context of this rapidly changing Ireland.

John Behan's Famine Ship invites the viewer to think and feel in unique ways about this tragic event. In that respect it is an important site in the production of new memories. It is clear that the memorial tries to bring people closer to the human realities of humiliation and despair, which lies at the heart of the Famine story. The real human suffering involved has been used and very often abused by those pushing a political agenda. It could be argued that the Famine was politicized from the very beginning. As Gemma Tipton (1999, 27) argues: 'all sorrow can be politicised, as all emotion is capable of manipulation. So whose sorrow is it that is being memorialised?' Behan in a sense tries to unravel the different layers of meaning that have accumulated over time in his efforts to present the visitor with a new way of seeing the Famine. In contrast the interpretation offered by Robert Shure's memorial in Boston is rooted in a particular myth. The difference in approach adopted by Behan and Shure reflects not only conflicting views of the Famine but also complex and diverse identities. It is a clear manifestation of a much larger issue, the contested nature of Irishness that exists between those living in Ireland and the diaspora overseas.

References

Daly, M. (1995), 'Why the Great Famine Got Forgotten in the Dark 1940s', *Sunday Tribune*, 22 January.

Gibbons, L. (1996), *Transformations in Irish Culture*, Cork University Press, Cork.

Gillis, J.R. (1994), *Commemorations: The Politics of National Identity*, Princeton University Press, Princeton.

Gray, P. and Oliver, K. (2001), 'The Memory of Catastrophe', *History Today*, 51, 2, pp. 9–15.

Howells, R. (1999), *The Myth of the Titanic*, Macmillan, London.

Humphreys, J. (1998), 'Famine Initiative Seen as Offensive', *Irish Times*, 30 October.

Johnson, N. (1995), 'Cast in Stone: Monuments, Geography and Nationalism', *Environment and Planning D: Society and Space*, 13, pp. 51–65.

Kearney, R. (1992), *Visions of Europe*, Wolfhound Press, Dublin

Kiberd, D. (2000), *Irish Classics*, Granta, London.

Larkin, E. (1998), 'Myths, Revision and the Writing of Irish History', *New Hibernia Review*, 2, 2, pp. 57–70.

Longley, E. (2001), 'Northern Ireland: Commemoration, Elegy, Forgetting', in McBride, I. (ed.), *History and Memory in Modern Ireland*, Cambridge University Press, Cambridge, pp. 223–253.

Mac Atasney, G. (1997), *'The Dreadful Visitation': The Famine in Lurgan/ Portadown*, Pale Publications, Belfast.

Murphy, J.A. (1995), *The College: A History of Queen's/University College Cork, 1844–1995*, Cork University Press, Cork.

Ó Gráda, C. (1999), *Black '47 and Beyond*, Princeton University Press, New Jersey.

O'Connor, J. (1995), *The Workhouses of Ireland: The Fate of Ireland's Poor*, Anvil Books, Dublin.

O'Toole, F. (1997), *Ex-Isle of Erin*, New Island Books, Dublin.

O'Toole, F. (1998), 'Turning the Famine into Corporate Celebration', *Irish Times*, 16 October.

Schama, S. (2001), *A History of Britain III: The Fate of Empire 1776–2001*, BBC Books, London.

Tipton, G. (1999), 'Lest They Fade', *CIRCA*, 89, p. 27.

Chapter 5

'Fostered To Trouble the Next Generation': Contesting the Ownership of the Martyrs Commemoration Ritual in Manchester 1888–1921

Mervyn Busteed[1]

Introduction

> Another remarkable contingent were a string of boys, apparently from some Roman Catholic school, nearly all of whom wore green neckties or green woollen comforters, and green badges on their arms. To an outside observer the reflection was irresistible that in these youthful processionists the story of recent Fenians was being fostered to trouble the next generation. (*Manchester Examiner and Times*, 2 December 1867)

On Saturday 23 November 1867 three young Irishmen were executed in Salford, England.[2] The executions followed the rescue of Thomas Kelly and Timothy Deasy of the Irish Republican or Fenian Brotherhood (I.R.B.) from a police van in Manchester on 18 September, during which police sergeant Brett was shot. Within a remarkably short time the three men had been transformed into almost secular saints as the 'Manchester Martyrs'. Commemoration of the event became a key date in the calendar of nationalists in Ireland and throughout the Irish diaspora, lasting well into the twentieth century and making it 'the longest lasting and most consistently observed political anniversary in the Irish nationalist tradition' (Owens, 1999, 32; McGee, 2001). In Manchester, as elsewhere, the commemorations quickly took on a routine, which crystallized into ritual. Like all ritual events, this annual

---

1   I would like to thank Manchester Geographical Society for financial support, the staff of the Local Studies Unit and Social Science Department, Manchester Central Library, Colindale Newspaper Library, National Library of Ireland, Ian Gavin, Andrew Gregg and Liam McLoughlin.

2   William Philip Allen, Michael Larkin and Michael O'Brien. Edward O'Meagher Condon was also sentenced to death but as an American citizen had his sentence commuted to life imprisonment. He was later released on condition he left British territory. His cry of 'God Save Ireland' when sentenced inspired the song of the same title, see Glynn, 1967.

commemoration seemed fixed in structure and meaning, but in reality was selective and flexible, dwelling only on chosen aspects of the happenings and personalities it claimed to commemorate. As with all commemorative events, it was used to legitimize contemporary groups and their agendas. The ownership of the event was regularly contested between moderate and militant Irish nationalists. This chapter will discuss the nature of commemoration and ritual and, using newspapers of varied outlook, will analyze the struggle for ownership of the commemoration ritual in Manchester between 1888 and 1921.

Contested Memory and Ritual

'For national communities, as for individuals, there can be no sense of identity without remembering' (McBride, 2001, 1). This sense of group remembering has been termed 'social memory', by which is meant the understanding and interpretation of events 'within a framework structured by the larger group' (Jarman, 1997, 6). From the earliest phase in the evolution of modern nationalism, the past has been a resource for nation builders. The romantic literary nationalists of the first half of the nineteenth century rescued, renewed or simply invented the oral and written traditions of their people and presented them as examples of the unique genius of a distinct group. In the second half of the century there was an accompanying trend to express the collective memory in monuments, processions and rituals in public space (Johnson, 1995). All such acts and artefacts of remembrance are prompts towards the construction of a consciously selective group memory.

Commemorative ceremonies are invariably endowed with an aura of respect verging on awe. In reality, commemoration is a flexible process subject to perpetual revision in response to the needs of the times. Features previously neglected or suppressed may be freshly emphasized and characters and happenings previously venerated may be downgraded. The purpose of such selective, shifting group remembering is to seek historical justification for current political attitudes and practices. Group memory is therefore a fluid, flexible construct subject to constant renewal and this means that the interrelationship between past and present is a complex one: 'It is the desires and aspirations of the present that shape our views of the past, while at the same time those present aspirations are partly formed by our understanding of the past' (Jarman, 1997, 5).

Ritual is an almost universal means of group commemoration. It has been described as a formalized, rule bound, structured and repetitive activity of a symbolic character restricted to specific times and places, which focuses the attention of participants and observers on 'objects of thought and feeling which they hold to be of special significance' (Lukes, 1977, 54; Kong and Yeoh, 1997). The compelling power of ritual lies in its multi-dimensional nature, which enables it to function at several levels. Ritual combines action and statement. Participants must follow a formalized sequence of physical movement and stillness at coordinated times and places. These moments of movement and stillness are full of meaning within the context of the organizing group. Ritual also appeals because it reaches out to involve both participant and observer until the distinction is blurred. The sheer performative spectacle of people in synchronized movement, possibly in distinctive dress, bearing symbolic colours or artefacts, accompanied by music, singing, chanting or slogans, draws and holds attention over a period of time, bonds spectator to participant and creates a sense of involvement (Bryan, 2000).

A similar duality of separation and involvement applies to the relationship of ritual to daily life. On the surface, ritual would seem to be the ultimate expression of liminality. The routinized nature of the participants, their dress, demeanour, movement, the almost liturgical nature of the language employed and the specific times of place and performance combine to set it apart. Yet such constant repetition of these widely recognized patterns makes them an integral part of society's life patterns. Moreover, if the ritual is on a large scale, then it requires resources, organization and rehearsal. It will also generate widespread public anticipation and excitement and will be the subject of retrospective discussion and comparison with previous events. Large-scale ritual is therefore characterized by both liminality and porosity, with its impact infiltrating into society both before and after the event (Jarman, 1997).

A final ambiguity of ritual, and one central to this discussion, is its combination of rigidity and flexibility. Part of the appeal undoubtedly lies in the constant repetition of formalized acts, generating a sense of order, continuity and legitimacy. However, collective appreciation of these features is not the same thing as common interpretation of the messages they transmit. Large-scale, regularly repeated ritual in particular can bear multiple meanings. It may mean different things to different groups at the same time. In addition, interpretations may alter with time, depending on the socio-economic and political context. It is this combination of constancy of form with flexibility and adaptability of meaning which lies behind its enduring strength and appeal (Bryan, 2000).

In the case of the ritual surrounding the commemoration of the Manchester martyrs in Manchester, the commemorative event has lasted into the early twenty-first century and, from quite early on, its ownership and significance have been the subject of robust dispute between the strands of Irish nationalism competing for the loyalties of the Irish migrant population of the city.

The Irish in Manchester

By 1861 there were 52,000 Irish-born people in Manchester, 15.2 per cent of the total population. Thereafter it declined slowly, until 1921 when it had fallen to 2.4 per cent, although by then there was also a considerable population of Irish descent (Fitzpatrick, 1989). The generally slender economic resources of the migrants meant they were concentrated in the poorer working class districts, with a tendency towards residential clustering in particular neighbourhoods (Busteed, 2000).

In 1857 James Stephens founded the I.R.B. in Dublin and, by 1864–5 there were groups in all British cities, with Lancashire the best organized district (Lowe, 1989).

It was from amongst the Manchester Fenians that the rescue and flight of Kelly and Deasy were organized on 18 September 1867. Demonstrations of sympathy for the executed men were held in Manchester on 24 November and 1 December, but there were no public commemorations for 20 years. When they did become established they were under very different auspices.

Sources

In the discussion of historic crowd phenomena such as public demonstrations and rituals a recurring problem is the dearth of records left by participants (Stevenson, 1979). Consequently, discussion must be based on material from observers of varying sympathies. In the present study the data bank is derived from newspapers. Clearly there are problems with bias, particularly where, as here, matters are politically contentious. However, the overt nature of the partisan stance, along with the variety of newspapers can render them a valuable source if used with caution (Harrison, 1988).

Four newspapers of varying attitudes towards the Irish are the chief data source. The *Manchester Guardian* (MG) during the period under discussion shifted its stance on Irish affairs. Originally supercilious and disdainful towards Irish Catholics and their politics, under the editorship of C.P. Scott (1872–1929) it became increasingly sympathetic towards Irish home rule. The *Manchester Evening News* (MEN) was owned by the same company and generally followed a diluted version of the same line. The *Manchester Courier* (MC) by contrast was always robustly anti-Catholic, anti-Irish and opposed to home rule. The weekly *Catholic Herald* (CH) provides a salutary counter to these three very English middle class dailies. Originally aimed at the London Catholic population, it quickly developed provincial editions and did not confine itself to purely church affairs, but carried extensive coverage of public issues as they affected Catholics and strongly supported Irish home rule.

Devising and Adapting a Commemorative Ritual

Public commemoration of the Manchester martyrs quickly became established as the main commemorative event in the Irish nationalist calendar, but there is no record of such a public ritual in Manchester itself until 1888. In the intervening 20 years commemoration took the form of requiem Mass in Manchester and Salford churches (*The Nation*, 30 November 1872).[3] Between 1888 and 1921, however, there are records of public commemorations in 24 of the 33 years and by the early 1890s the ritual had become a customary part of Manchester Irish life (Fielding, 1993).

The event had a definite ritualistic structure. The date normally chosen was the Sunday closest to the execution anniversary of 23 November. Participants assembled at about 9.30 a.m. Between 1888 and 1907 the assembly point was New Cross at

---

3   This paper was published in Dublin and was strongly nationalist.

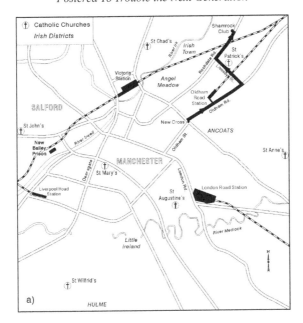

Map 5.1a   Processional Routes Used in the Manchester Martyrs
            Commemoration Rituals, 1888–1907

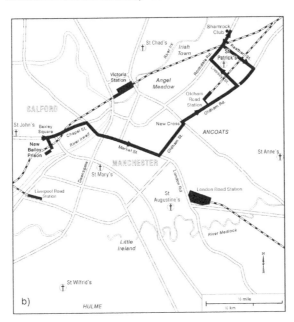

Map 5.1b   Processional Routes Used in the Manchester Martyrs
            Commemoration Rituals, 1908–1921

the junction of Oldham Street and Great Ancoats Street north of the city centre on the fringes of the largest Irish districts (Map 5.1a). A procession was formed which walked to nearby St. Patrick's Church for the celebration of Mass. The procession then reformed and walked to either the parochial hall or the nearby Shamrock Club, meeting place of the local branch of the Home Rule Movement. There a public meeting was held, at which resolutions on Irish affairs were debated and an address given by an invited guest, usually a political activist. This normally fell into two parts, a tribute to the dead heroes and a commentary on Irish current affairs, especially the campaign for home rule. By 1891 the event had become so well established that one commentator referred to 'the usual demonstration' (*MEN*, 23 November 1891).

Analysis reveals many of the characteristics of ritual. There was a definite bonding function at several levels. Speakers made strenuous efforts to relate 1867 to contemporary concerns. One approach was to encourage identification with the deceased by stressing the human pathos of their situation. In 1896 a speaker drew attention to Larkin's family and asked, 'what was to become of his three helpless, fatherless children? [...] Who dandle them on his knee? Who fondle them in his arms and nurse them when ill?' (*CH*, 27 November 1896). The Catholic Church was anxious to claim them as good Catholics, political speakers lauded them as patriots and others conflated the two, as in 1888 when it was claimed: 'They died as true Catholics as Irish men should, professing their faith and with a prayer for their country on their lips' (*CH*, 23 November 1888).

Every significant Irish organization in the city was represented, though the Irish National Foresters, a nationalistic insurance association, and the Irish National League (United Irish League after 1900), the political arm of the Irish home rule movement in Great Britain, were the most stalwart. By 1912 the Ancient Order of Hibernians, another nationalistic mutual aid organization, were present and in the same year the Gaelic League and the Gaelic Athletic Association were represented (*CH*, 30 November 1912). In 1919 Sinn Féin took part for the first time (*MEN*, 24 November 1919).

Speakers at the public meeting urged listeners to stand together as Irish people. In November 1902 John Murphy, Nationalist MP for Kerry East exhorted Irish migrants to 'seek out in all the big English cities where their lot may be cast, every member of their race and bind them together in one united bond of affection. Let no man cast aside the lowliest or meanest Irish man or woman that breathed in Manchester that day' (*CH*, 28 November 1902). But this sense of fellowship was meant to extend beyond Manchester. Speakers constantly reminded listeners that 'they must try to direct their enthusiasm towards the benefit of Ireland' (*MC*, 24 November 1902). As if to compensate for their migrant minority status as 'exiled sons of Erin' (*CH*, 27 November 1896) audiences were told that 'these demonstrations took place in Australia and other colonies, as well as in the towns of England and Ireland' (*MG*, 23 November 1908).

The ritual dimension was emphasized by music. At least one band always accompanied the procession and singing was often incorporated in the public meeting. The two pieces most frequently heard were 'The Dead March' and 'God

Save Ireland' composed in honour of the martyrs a few weeks after their execution and variously referred to as 'The National Anthem' (*CH*, 29 November 1889) and 'the national prayer' (*CH*, 6 December 1913). Some participants wore ceremonial dress or carried distinctive emblems, and on several occasions spectators dressed and behaved in a fashion blurring the divide between participant and observer. In 1902, for example, 'The large crowds who greeted the processionists *en route* wore green favours and shamrock devices, and showed their appreciation of the procession by lustily cheering as it moved along' (*CH*, 28 November 1902).

Flexibility is demonstrated by occasional modifications. It has already been noted how the burgeoning Gaelic League and GAA were incorporated. In November 1913 the event was postponed for one week because it clashed with the civic service for Daniel McCabe the city's first Catholic Lord Mayor. In 1915 wartime conditions forced cancellation and in 1916 there was a Mass and public meeting but no procession. There was no commemoration in 1918, possibly because the war had just ended and the first general election campaign for eight years, and a particularly important one for Ireland, was under way. In 1908 an almost entirely new route was adopted. Participants now assembled in Bexley Square, Salford, at 9.30 a.m. on the appropriate Sunday and proceeded along Chapel Street into New Bailey Street. Here at the place of the 1867 executions the procession paused, now a sacred site for the Manchester Irish, heads were uncovered, and the band played 'The Dead March' (*MG*, 23 November 1908). The procession then resumed its march to Mass at St. Patrick's, followed by the usual public meeting in either the parochial hall or the UIL club (Map 5.1b). In 1919 an entirely different route was followed. Mass was at St. Anne's, Fairfield in the morning and in the afternoon a procession marched to Moston cemetery, but this was an aberration. When in 1921 the first commemoration entirely under Sinn Féin auspices was held, it assembled in Bexley Square and followed the traditional route.

The Ritual Contested

Clearly, ritual commemoration ceremonies are flexible in form and meaning, so that within the organizing group 'rituals may be contested and fought over, praised and condemned and ultimately transformed in meaning' (Jarman, 1997, 11; Marston, 2002). Consequently, the varied elements within a community may come together to participate in a ritual whilst differing in their interpretations. Each celebration of a ritual is, therefore, a balancing act and behind the scenes there is an endemic tension. Actual control of the ritual can shift in response to outside events or long-term contextual changes. In such situations while the form of the ritual is retained, its meaning may be radically altered.

In the Manchester situation the basic axis of contestation was between what may be termed 'moderate' and 'militant' Irish nationalists. Moderate nationalists wished to see a measure of Irish home rule within the United Kingdom and Empire achieved through peaceful parliamentary methods. Their political vehicle was the Irish

National League, after 1900 the United Irish League, whose most notable leaders in the period were Charles Stewart Parnell and John Redmond. Militant nationalists aspired to an independent Irish republic to be achieved by force of arms. Their most enduring vehicle was the secretive, schismatic IRB, though from 1917 onwards the Sinn Féin party became increasingly significant.

Control of this commemoration was certainly worth contesting. The events of 1867 had rapidly assumed iconic status for Irish nationalists everywhere. Whatever their undoubted abilities as organizers of public spectacle, the rescue of Kelly and Deasy was one of the few occasions when Fenians could claim anything resembling a military success. Moreover, the three executed men were the first Irish nationalists tried and executed by British authorities for a political crime and therefore the first credible martyrs since Robert Emmet in 1803.

Throughout most of the period under discussion moderate nationalists controlled the commemoration ritual, though it will be shown that this control was never totally confident and was eventually successfully challenged. As already noted, there were processions of support for the executed men on the two Sundays immediately following the executions on 23 November 1867 and their disciplined nature strongly suggests they were under Fenian control (*MG*, 2 December 1867). However, when public commemoration became established in the late 1880s, moderate nationalists were in charge. They had to perform a complex balancing act. They were moderate people commemorating indisputably violent acts and they were aware of the strong popular sympathy for the patriotism of the Manchester martyrs, if not for their aims or methods. Moderate nationalists selectively utilized the events and personalities for their own political purposes, but were clearly vulnerable to challenge.

Analysis of the speeches at the public meetings reveals the tactics used to deflect, contain or restrain the more militant elements. One approach was to emphasize the need for unity amongst Irish nationalists and this may also have been a reaction to the bitter split which had occurred in the Irish parliamentary party in 1890 over the Parnell divorce case. In Manchester the local nationalist political organization divided but control of the commemoration ritual remained with anti-Parnell elements (Henderson, 2003). However, speeches at the public meeting were notable for their avoidance of the issue, possibly from a desire not to exacerbate the situation. They tended to focus on matters which all nationalists could support, such as the demand for an amnesty for Irish political prisoners, which was the subject of resolutions in 1893, 1896 and 1898. In 1896 the main speaker warned his audience to be on their guard against those who would introduce 'that dangerous reptile – dissension – into your ranks' and exhorted them to: 'Never forget that every act of dissension [...] is the greatest stumbling block to the progress of the cause of Erin, a scandal to those who would befriend her, and an occasion of mirth to her enemies' (*CH*, 27 November 1896). By the commemoration of November 1899 negotiations to reunite the party were almost complete and at the meeting it was reported that 'Referring to the follies of recent years, Councillor McCabe urged all Irishmen to forget the past, to unite together in Manchester, and throughout England, to join the Nationalist organization' (*CH*, 1 December 1899). But the appeals for unity did not cease, partly because the recent experience had been so deeply wounding and partly because there were ongoing stresses within the nationalist movement. At the 1908 commemoration John Muldoon, MP for Wicklow East, noted recent progress and drew the lesson that: 'Dissension had been the curse of Irish parties but they were now a strong and united party' (*MG*, 23 November 1908).

By the early twentieth century there was a process of cultural and social ferment under way in Irish society. As yet it was not strongly reflected on the political scene, but some of the more observant members of mainstream nationalism made efforts to accommodate these developments. It has already been noted how the GAA, the Gaelic League and the restructured AOH were co-opted into the commemoration ritual. The underlying concern was that these new developments could distract the home rule movement. This was well illustrated in 1913 when the Dublin strike and lockout was at its height. The union leader James Larkin had addressed packed and enthusiastic meetings in Manchester, the latest on 16 November, raising the prospect of class divisions (Herbert, 2001). At the martyrs commemoration meeting on 1 December the main speaker was the Secretary of the United Irish League. It was reported that: 'With reference to the Dublin trouble, Mr. Donovan asked the people not to allow any subsidiary interest to jeopardize their ideal – national self-government. (Applause)' (*CH*, 6 December 1913).

Until 1916 moderate nationalist control seemed secure, yet analysis suggests that at least some organizers and speakers were aware of the ongoing appeal of the more militant tradition. Its latent strength and the political cross currents within the Manchester Irish were revealed on 27 November 1898 when the foundation stone of a monument to the martyrs was laid in Moston cemetery in north Manchester. The leaders of Manchester nationalism had been present in strength at the traditional martyrs' commemoration the previous Sunday but were conspicuous by their absence on this occasion. However, some notable advanced nationalists were present including an elderly James Stephens and the actress Maud Gonne who gave a rousing address at a public meeting. An indication of the strength of feeling the memory of 1867 could still invoke is given by the comment that the numbers attending the laying of the foundation stone were 'an enormous crowd' (*Manchester Evening Chronicle*, 28 November 1898).

The point was underlined by the unveiling ceremony for the completed monument on 5 August 1900. Once again, none of the leaders of mainstream Manchester Irish nationalism were present. At the meeting following the ceremony both political aims and methods were made clear. Seamus Barrett, President of Manchester GAA, declared: 'Their one aim was the absolute independence of Ireland' (*CH*, 10 August 1900). John Daly, IRB member and now mayor of Limerick, who had performed the unveiling, argued that 'the spirit created by the Martyrs' blood was strangled by a slavish sentiment, a sentiment that rose and was promulgated by selfish politicians who would barter the rights of Ireland away for concessions […] It had been well if Irishmen had not listened to the "deluderings" of Parliamentarians' (*MC*, 6 August 1900).

Moderate nationalists deployed a variety of rhetorical devices to cope with such sentiments. In 1902 John Murphy MP acknowledged that there were 'many movements [...] which tried to fulfil the martyrs' aims, but appealed to men and women of advanced opinions [...] that when they could not do the best thing, to do the next best thing and join the national organization' (*CH*, 28 November 1902). Other speakers adopted the fiercely militant rhetoric then characteristic of Irish political discourse (Jackson, 1999). In 1912 William Redmond, son of the parliamentary leader, declared: 'He offered no apology [...] for the aims and ambitions of the Irish Fenians, but solemnly declared that had he been living at the time he would not have hesitated to have fought for his native country' (*MG*, 25 November 1912).

More often, speakers went to some pains to distinguish between Fenian aims and methods. In 1909 John Cullinan, MP for Tipperary, declared that although '[Fenian] methods were different from the methods of today, the objects were the same' (*MG*, 22 November 1909), conveniently glossing over Fenian separatism. Others tried to reconcile past and present by arguing that whilst they admired the aims and personal qualities of the martyrs, the context had changed. In 1902 Dan McCabe used this argument: 'He sincerely hoped that the spirit which actuated the Manchester Martyrs, a spirit of love sacrifice for Ireland, would go on year by year in finding fresh and ardent adherents. Different methods might be resorted to according to the time and circumstances' (*CH*, 28 November 1902). Still others made audacious efforts to claim continuity and even convergence between the Fenians and the Irish parliamentary party. In 1902 John Murphy had gone on to argue, 'What the Fenian movement was in the past, the United Ireland League movement was now, for it was animated by the same principles and had the same object in view. Fenianism meant love of Ireland, hatred for its enemies and a desire to set the country free. Parliamentarian though he was, he had no hesitation in saying, they and he would take pride in being Fenians still. (Cheers)' (*MC*, 24 November 1902).

In September 1909 the city's constitutional nationalists played a significant card when Edward O'Meagher Condon returned to the city, visiting the sites associated with 1867 and taking part in a large and enthusiastic public meeting on 26 September. Though resident in the USA he showed he was attuned to the nuances of Irish nationalist politics. He cited the accomplishments of the parliamentary party and drew attention to the fact that: 'On his side of the world there was practical unanimity in support of the claims of Ireland and of the Party she put forward ... [justice] could only be achieved by loyal support of the Party they had elected to support them in the British Parliament. (Cheers)' (*CH*, 2 October 1909).

By late 1911 British political circumstances made home rule seem imminent. At the meeting on 26 November 1911 Thomas Scanlon, MP for Sligo North made a remarkable statement which, whilst perhaps calculated to reassure the unenthusiastic Liberal government expected to sponsor home rule, nevertheless represented a considerable departure from the fiery rhetoric heard at this event in previous years. 'The people of Ireland did not want separation; they wanted the right of Irishmen on their own soil to manage purely Irish affairs – the same rights as now enjoyed by the colonies. They wanted a measure that would be safe for the Imperial Parliament and safe for the Empire' (*MG*, 27 November 1911). By November 1913, when the Home Rule Bill had passed the Commons, J.T. Donovan informed the meeting that 'they would be celebrating next year the victory of the principle of the Irish people, on their own soil, along their own lines, according to their own ideas, to manage and control the internal and domestic affairs of their own country. (Applause) [...] Home rule was now only a matter of weeks' (*CH*, 6 December 1903).

Nationalist confidence climaxed in November 1914, when a Home Rule Act was passed, though suspended for the duration of the war. Redmond, the nationalist leader, had declared full support for the British war effort. It was possibly the music which best expressed the confidence and the ironies on display. The *Manchester Guardian*, with the headline 'Irish Loyalty at the Annual Celebration' noted how the ritual had been adapted to the circumstances: 'In the usual ceremonies there was one significant change – the playing by the band of the national anthems of all the allies, and of "It's A Long Way To Tipperary"'(*MG*, 23 November 1914). 'At the close of the public meeting the choir boys of St. Patrick's sang the "Marseillaise" and Irish airs' (*CH*, 28 November 1914). But even amidst this triumphalism some noticed there was a small cloud on the horizon, since speakers found it necessary to denounce: 'the mischievous policy of the small anti-Redmond faction' (*MG*, 23 November 1914).

The nationalists of Manchester were never again to be so confident. As already noted, no commemoration was held in 1915 and only a Mass and public meeting in 1916. The atmosphere on this occasion was very different. It was noted that the events 'were of a very quiet character [...] the attendance, owing to so many men being with the colours, was small' (*CH*, 2 December 1916). The speeches were equally low key and indeed somewhat hesitant in view of the rising of Easter 1916 and its aftermath, which were not directly mentioned. The homily at Mass focused on the tradition of prayers for the dead without mention of specific people. Some of the speeches were almost apologetic. The chairman went to some lengths to explain: 'They were not holding that commemoration in any sense of menace or defiance, nor did they wish to obstruct in the slightest those who were prosecuting the war to a successful conclusion, because they had an interest in it' (*CH*, 2 December 1916). There were constant appeals to support the parliamentary party and its leadership: 'Mr. Redmond had given long service to the Irish cause, and the Irish party were men to be reckoned with. They would do credit to any country [...] the policy of Mr. Redmond was the only one by which they would get back freedom for Ireland' (*CH*, 2 December 1916).

The following year was the 50th anniversary of the executions and the full ritual of procession, Mass and public meeting was restored. Once again there was an air of ambiguity about the proceedings. The homily expounded traditional Catholic teaching on conditions justifying insurrection without offering judgement on the 1916 rising. The chairman managed to be both defiant and defensive: '...for the past two years they had not held a demonstration, but this year it was considered desirable, even necessary, to do so. They made no apology to anyone for holding it.

But at the same time they did not hold it in any spirit of bombast or to aggravate the feelings of those amongst whom they lived' (*CH*, 1 December 1917).

By 1919 it was clear that the nationalists were losing control of the event. Many of its defining features were missing. The commemoration began with Mass in the morning of 23 November in St. Anne's church, Fairfield. In the afternoon a procession assembled in Stevenson Square. For the first time, Sinn Féin was represented. The Gaelic League and the GAA were also represented, but there is no mention of the UIL, Foresters, or AOH. A new route and destination were used: 'The demonstrators, among whom were a good many women with flags and banners, and young men carrying hurley sticks, formed into procession in the Square and headed by Irish pipers marched to the Roman Catholic cemetery at Moston ... a large crowd had assembled, the Rosary was recited in Gaelic' (*MEN*, 24 November 1919).

In 1920 ownership had clearly passed to Sinn Féin. Originally it was intended to incorporate the commemoration into a convention of the Irish Self Determination League (ISDL), the British branch of Sinn Féin, to be held on 21–23 November. This was a notably tense time in Ireland, with the War of Independence entering a particularly grim phase. In mid-November Manchester police had seized documents leading them to believe there was an IRA plan to blow up one of the city's main electricity generating stations (*MEN*, 25 November 1920). The result was a ban on all the proceedings. In fact the ISDL meeting went ahead at a new venue and there was also a 'wildly enthusiastic' public meeting on Sunday 23rd in St.Patrick's Hall at which one of the speakers was Sean Milroy, director of organization (*MG*, 29 November 1920). He declared: 'England was no longer dealing with a mob of politicians, but an organized nation, and the Irish nation would never again assume the crouching mendicant air' (*CH*, 4 December 1920). The sole UIL contribution was a rather wan letter to the local press distancing them from the alleged bomb plot: 'the United Irish League, the older organization of Irishmen in the district, now as always, stands for constitutional measures of reform, and disassociates itself from any connection with the so called plot' (*MEN*, 26 November 1920).

In 1921 all the elements of the traditional commemoration were restored, beginning with the assembly at Bexley Square at 9.30 a.m. on Sunday 20 November. After Mass at St. Patrick's the main speaker at the meeting in the parochial hall vigorously asserted Ireland's right to self government. But, however identical the ritual, the meaning was rather different. The speaker was Sean Milroy, now Sinn Féin TD for Cavan-Monaghan, and his appeal was to support the Sinn Féin delegation then negotiating with the British government in London. Perhaps the transition and the linkage were best summed up by the newspaper headline 'An Old "Martyrdom" and a Modern Cause' and the comment: 'To the Manchester Martyrs Allen, Larkin and O'Brien, whose memory has been honoured in this way for so many years, have now been added the names of Kevin Barry, Thomas MacCurtain, Terence MacSwiney, as well as the Irishmen who fell in Easter week, 1916 and since' (*MG*, 21 November 1921). The new wine of Sinn Féin sat comfortably in the old wineskin of the traditional commemoration ritual.

Conclusion

Between 1868 and 1921 Manchester's commemoration ceremonies for its martyrs displayed marked ritualistic features. Superficially, there was an air of venerable rigid immutability, whereas in reality every aspect of the event was sensitive to contemporary political conditions and open to negotiation and adjustment. To some extent these shifts reflected events in Ireland, but more often they reflected the varying fortunes of the movement for some form of Irish self-government. In its earliest manifestation the commemoration may have been intended as an act of resistance to British hegemony. By the early 1890s however, whilst it retained its bonding function for the local Irish population, it seems to have become a broadly accepted feature of the city's cultural landscape.

Yet from quite early on it had become an event whose ownership and meaning were contested amongst Manchester's Irish population. Moderate and militant nationalists presented differing versions of Ireland's historical narrative and political future and whoever controlled this event could use it to validate and express their viewpoint. Until 1917 moderate nationalists maintained ownership, but the ceremonies surrounding the martyrs' monument in 1898 and 1900 and the rhetorical gymnastics of the speakers at the public meetings reveal an uneasy awareness of the ongoing appeal of the physical force movement. By 1921 control had passed totally into the hands of more militant nationalists, but whilst the message conveyed was now separatism achieved through armed insurrection, the structure of the event was identical in every respect to the long established pattern.

The case of the Manchester martyrs illustrates the fact that group memory is a malleable construction and a valuable historical resource for contemporary political purposes. So long as a commemorative ritual can be appropriated and adapted, then it will survive and help in the ongoing construction and renewal of memory and identity and competing groups will find it worthwhile to contest its meaning and ownership.

References

Bryan, D. (2000), *Orange Parades: The Politics of Ritual, Tradition and Control*, Pluto Press, London.

Busteed, M.A. (2000), 'Little Islands of Erin: Irish Settlement and Identities in Mid-Nineteenth Century Manchester', in MacRaild, D. (ed.), *The Great Famine and Beyond: Irish Migrants in Britain in the Nineteenth and Twentieth Centuries*, Irish Academic Press, Dublin, pp. 94–127.

Fielding, S. (1993), *Class and Ethnicity: Irish Catholics in England 1880–1939*, Open University Press, Buckingham.

Fitzpatrick, D. (1989), 'A Curious Middle Place', in Swift, R. and Gilley, S. (eds), *The Irish in Britain 1815–1939*, Pinter Publications, London, pp. 10–59.

Glynn, A. (1967), *High Upon the Gallows Tree*, Anvil, Tralee.

Harrison, M. (1988), *Crowds and History: Mass Phenomena in English Towns 1790–1835*, Cambridge University Press, Cambridge.

Henderson, P. (2003), 'An Irish Great Grandfather', *Manchester Genealogist*, 39, 1, pp. 31–8.

Herbert, M. (2001), *The Wearing of the Green: The Political History of the Irish in Manchester*, IBRG, London.

Jackson, A. (1999), *Ireland 1798–1998*, Blackwell, Oxford.

Jarman, N. (1997), *Material Conflicts: Parades and Visual Displays in Northern Ireland*, Berg, Oxford.

Johnson, N. (1995), 'Cast in Stone: Monuments, Geography and Nationalism', *Environment and Planning D: Society and Space*, 13, pp. 51–65.

Kong, L. and Yeoh, B.S.A. (1997), 'The Construction of National Identity through the Production of Ritual and Spectacle: An Analysis of National Day Parades in Singapore', *Political Geography*, 16, 3, pp. 213–40.

Lowe, W.J. (1989), *The Irish in Mid-Victorian Lancashire: The Making of a Working Class Community*, Peter Lang, New York.

Lukes, S. (1977), *Essays in Social Theory*, Columbia University Press, New York.

Marston, S.A. (2002), 'Marking Difference: Conflict over Irish Identity in the New York St. Patrick's Day Parade', *Political Geography*, 21, 3, pp. 373–92.

McBride, I. (2001), 'Memory and National Identity in Modern Ireland', in McBride, I. (ed.), *History and Memory in Modern Ireland*, Cambridge University Press, Cambridge, pp. 1–42.

McGee, O. (2001), 'God Save Ireland: Manchester Martyrs Demonstrations in Dublin 1867–1916', *Éire-Ireland*, 36, pp. 52–73.

Owens, G. (1999), 'Constructing the Martyrs: The Manchester Executions and the Nationalist Imagination', in McBride, L. (ed.), *Images, Icons and the Irish Nationalist Imagination, 1867–1925*, Four Courts Press, Dublin, pp. 18–36.

Stevenson, S. (1979), *Popular Disturbances in England 1700–1870*, Longman, London.

PART II
The Politics of Heritage and the Cultural Landscape

Chapter 6

Changing Conceptions of Heritage and Landscape

Paul Claval

The contemporary interest in landscapes is often linked to their role as sites of memory. They appear as key elements in the building and preservation of collective identities, yet it would seem that Western societies are today experiencing a crisis of these same identities. This chapter explores how the conception of culture and landscape has changed historically, along with the role and significance of geography. It argues that despite this flux, landscape remains an important medium through which to interrogate the construction of identity and the politicization of space.

Culture in Historical Perspective

In the English speaking world, social scientists generally categorize culture as being either high or low brow, the French equivalents of which – *culture populaire* and *culture élitaire* – perhaps have richer connotations. It is useful to bear this in mind in the context of this chapter which identifies distinct differences between historical, vernacular culture and more contemporary, political or constructed understandings of culture. Jackson (1994) in *A Sense of Time, a Sense of Place* focuses on the former, highlighting the often-disregarded importance of vernacular culture in identity fomation. For Jackson, what is significant is the fact that the transmission of this culture relied mainly on direct imitation, observation and words. It meant that the group from which this inheritance came was essentially that of the immediate community and surroundings, and the reading of these landscapes by the local population became more important than its official interpretation. As long as cultural transmission was based on face to face relations, observation, imitation and the exchange of words, individuals could experience a sense of continuity and connection between their past and future.

Within such an orally-based society, time was structured in a simple way and history was understood within the timespan covered by the memories of its living members. As a result, the past was never completely a foreign country, since those who witnessed it were still alive. All that happened before the birth of the oldest members of the community was immemorial – it could not be known directly. The answers to the questions on the origin of the group, the significance of its presence in this World, its environment and its organization were provided by the myths which served as substitutes for memory in the time that could not be directly recalled. As a result, it could be argued that individuals and societies had a clear sense of their place, and of who they were. For them, there was no need to explore the past in order to discover their identities, since the time that counted for them was directly experienced as a living reality.

In contrast, the emergence and development of high culture was made possible not through face-to-face relations and oral exchange, but through the significance of the written word. Writing transforms culture into objective memories and such a transformation has both temporal and spatial consequences. Regarding time, the elements transferred to younger generations cease to be limited to what is alive in the minds of their parents and neighbours. Older components are incorporated, giving culture a new historical perspective. On the spatial scale, there is a similar expansion in spheres of interest as distant places become incorporated into the story of local societies. High culture is more specifically centred on intellectual knowledge and values and thus became useful in the building of power systems in the nineteenth century. Out of the existence of perceived geographical and historical bonds, it was possible to begin building great systems of power and political organization. As a result, the concept of nation became a meaningful category for social, historical and political analysis giving rise to issues surrounding the formation of collective identities.

In high culture, participation within a group results from the sharing of an objective and dead memory, the memory encapsulated in texts. Myths, and directly remembered events, are replaced by revealed or interpreted philosophies of history, suggesting that collective identities thus have a historical basis. Identity thus becomes intimately connected with these constructed memories and history becomes a useful tool, helping both construct and chart an understanding of the progress of particular societies. In the nineteenth century the integration of individuals into political communities built on the delegation of divine or rational authority to a king or emperor resulted from their perceived personal allegiance to him. In *Ancien Régime* France, people were first and fundamentally the subjects of the King of France; it was for that reason they considered themselves French. By the end of the nineteenth and the beginning of the twentieth century, people generally participated in two main levels of identities: the local one, learnt through the living memory of vernacular cultures, and the national one, produced by history and taught through the school system.

The Contemporary Mutations of Culture: History, Memory and Heritage

Today, however, this division between tradition passed down through the vernacular components of cultures, and history as a political narrative appears to be diminishing, primarily one could argue due to the role of the modern media. Face to face relations through which gestures, attitudes and know-how were transferred from one generation to the next on a local basis now compete with external influences. With the modernization of techniques, such as the rise in the availability of satellite television, 24-hour news channels and the worldwide web, traditional know-how has been replaced. Vernacular cultures appear to be increasingly considered as localized components of global cultures and direct knowledge of the way in which daily life has been produced has disappeared.

The second major event in the evolution of the relations of Western societies to their past and the building of their identities, resulted from the death of the philosophies of history and progress. Everyone is aware of the new possibilities that science has opened in daily life, but the proportion of those who are afraid of the new potential for destruction that it creates continues to escalate. People have lost faith in the capacity of technical progress to provide happiness and thus, the whole ideological basis for modernity, as developed in the Western World since the time of the Enlightenment, has collapsed. The nation-state, which appeared as the tool offered to each particular group to achieve in this World its own version of paradise, has lost its main *raison d'être*.

This crisis of modern ideologies is a major problem for the contemporary World, and more specifically for Western civilization, because it is there that the transformation of vernacular cultures has been the deepest and where national identities have played the most significant role. People react to such a situation and develop new strategies to preserve memory and create identity:

- In order to keep alive the vernacular forms of identities, people try to preserve the material environments of the past, giving newfound importance to the role of heritage.
- Local, regional and national identities are often redefined. Instead of being the expression of social groups trying to organize their life and environment through political action, they are often perceived as ethnic entities. Their 'truth' is based on the cultural interpretations given by past generations to particular places.
- For many people, especially outside the Western World, the solution to the identity problem is to look towards the past to build societies that are capable of solving contemporary problems, often giving rise to extreme fundamentalism.
- A fourth solution is to build new ideologies. In order to look credible, they have to rely on other forms of history. Multiculturalism is based on an analysis of human evolution which has ceased to be centred on individuals and nations, but is focused on cultural communities. Ecologism is built on a conception of evolution which has ceased to be organized around societies: what is at stake is nature. The duty of people is to provide the Earth with an opportunity to evolve naturally with minimal human interference.

The Evolving Conception of Heritage

For geographers interested in the spatial patterning of social life and the symbolic imprint of social groups in the landscape, this change in attitude towards heritage is highly significant. The idea of heritage emerged at a time when religious or metaphysical beliefs ceased to exist as the main bases for collective values, social life and political organization. It was the time when nations became the only justification for the existence of states, the era of the nation-state. In Western Europe, where the existing states coincided approximately with national groups, the heritage that was preserved consisted mainly of the palaces and castles of former kings, the cathedrals or monasteries that best illustrated the religious faith upon which their action relied, and the castles and mansions of the aristocracy which helped them. In Central and Eastern Europe, where the existing States did not coincide with national groups, the heritage preserved was mainly that of the people themselves. Initially the emphasis was placed on its non-material components, such as language, popular poetry, popular music. However by the mid-nineteenth century, the material bases of daily life, such as, tools, artefacts and houses, began to be incorporated into the idea of heritage. Today the notion of heritage is much broader than in the past, and this has profound spatial implications. New values and meanings are now being ascribed to particular landscapes, many of which previously were not considered of particular significance.

Landscapes, Memory and Identities in Historical Perspective

As has been suggested above, the place of or importance ascribed to landscape as experienced by Western civilizations, and elsewhere, has utterly changed since the beginning of modernity. I shall organize this analysis in two parts: first, the period before the mid-twentieth century, and secondly, the period from the 1950s on. While many of the people who tried to build the new national ideologies that were called for in the eighteenth and nineteenth centuries were conscious of their role in local life, this did not appear as an essential component in the enterprise they were undertaking. Yet, the essayists, painters, artists and historians who were mainly responsible for the construction of the idea of nation stressed the role of landscapes for various reasons (Appleton, 1996).

Some landscapes were highlighted because they were considered as essential in the building of the nation and the shaping of its culture. This was most obvious in Switzerland where the Swiss character was forged by the daily confrontation with the difficult mountainous environment of the Alps. Lunn (1963) suggests that the wonderful scenery gave those who inhabited it an opportunity to develop a sense of dignity and grandeur. Similarly, in Denmark, the heath of Western Jutland played a key role in the development of modern Danish national consciousness in the nineteenth century, as analysed by Kenneth Olwig (1984). This landscape was associated with the kind of Gothic imagination or romantic sensibility which prevailed at that time. Since it was a poor environment, its reclamation from the mid nineteenth century appeared as a major national achievement. And in Britain, Cosgrove and Daniels (1988) have suggested that landscapes offered an image of harmony between natural and human forces, and of a hierarchical society where class conflicts had remained benign. At a time when the strength of Britain was linked with its collieries, black countries and industries, the emphasis on the kind of landscape painted by Constable erased all forms of violence, injustice and crude power relations from the national consciousness.

Beyond Europe, it is clear that similar kinds of imaginings were also occurring. In the United States for example, people appeared more sensitive to the significance of mountain landscapes, the Catskill mountains first and later the Rocky Mountains. The beauty of the big desert landscapes and the monumentality of the Colorado canyon were highly prized. Writers, painters and photographers participated in this valorization of the most dramatic American landscapes. In Canada, the Ontarian Group of Seven contributed much to the development of national identity in the first decades of the twentieth century through their painting of wild areas of the North, where the inherent grandeur and strength of Canadian nature were most evident. At the same time, painters in Quebec tried to develop the collective consciousness of the French-Canadian community, through their favourite theme of the *rang* – the long-field system, especially in the marginal areas where it signified a victory of man over a difficult nature (Lasserre, 1993).

Yet contrary to what might be expected, the exploitation of landscape for the construction of national identities was not, at the end of the eighteenth century and during most of the nineteenth century, a geographical enterprise. Geographers had not yet developed the tools they needed to cope scientifically with landscapes. As long as they considered their main task to be a description of the variety of the Earth's surface, they remained disengaged from ideological debates.

But for those who shared an environmentalist conception of the discipline, the situation was different. Herderian geographers of the early nineteenth century were more sensitive to the harmony between the genius of a people as expressed in its language, poetry of music and the country it inhabited, than to its landscapes. For the Darwinists, landscapes were even more significant as they channelled *man-milieu* relationships. The situation changed utterly during the last decade of the nineteenth century, when August Meitzen published his great study on the forms of rural landscapes in Europe (Meitzen, 1895). For him, each people created a special form of land organization and thus landscapes became markers of the ethnic or national genius of human groups.

In *Tableau de la Géographie de la France*, Paul Vidal de la Blache considered that the geographic personality of France – which gave it its unity and justified the limits of its territory – resulted from the very diversity of the country (Vidal de la Blache, 1903). What gave the French nation its specific character was the opportunity the country offered its inhabitants to build efficient territorial forms of organization out of the complementarity which existed at the local, regional or national scales. The *Tableau* provided the French nation with a scientific caution. But while the contribution of geographers to the earliest debates on spatial identities and the significance of landscape remained relatively limited, the contemporary situation is utterly different.

Landscapes and the Construction of Contemporary Identities

With the increased importance of the archivistic as opposed to oral form of memory, all landscapes, past or present, are transformed into potential forms of heritage. For those who struggle to maintain local identities, preventing the transformations that would deprive local populations of landscapes representative of their historical triumphs is crucial. Central to this is preservation, either of whole landscapes or parts of them worthy to be safeguarded. But this is not an easy enterprise and is in fact highly political. Those who try to promote the preservation of this new form of heritage, or the public administrations which are responsible for the application of new policies in this field, rely heavily on the landscape studies written by geographers in recent times on the organization of traditional and modern landscapes, and geographers are often asked to serve as experts in the research policy process.

So, does that mean that geographers fully agree with these new orientations? It is unlikely that this is the case and their studies simply serve to establish the dynamic nature of landscapes. Geographers often highlight the difficulties in conceiving of landscapes as rigid combinations of structural patterns. Instead they contextualize the debate, indicating the evolution of particular places in response to demographic and economic pressures, the legal system of inheritance, the interest in open air activities and sports. The idea of landscape preservation appears to most geographers as a dangerous mistake. They agree with the necessity of safeguarding a part of what has made past landscapes so interesting, varied and significant for the people who inhabited them, but know that it is dangerous to prevent economic innovation and impossible to prevent the modernization of at least a part of settlement patterns and transportation networks. Their idea is generally to focus the action of preservation on only a few aspects and areas.

The contribution of geographers, therefore, to both the analysis of the manner in which landscape was appropriated for political means in the nineteenth and early twentieth centuries, and contemporary debates around the preservation of vernacular landscapes, is essentially a critical one. The idea of landscapes, especially rural lansdcapes, as an expression of the genius of a particular ethnic or regional group does not withstand scientific scrutiny. The interpretation of landscapes published by Meitzen in 1895 has been severely damaged by Roger Dion's publication (Dion, 1946), and definitively condemned by Anelise Krenzlin's studies in the 1940s (Krenzlin, 1957). The criticism of the Vidalian idea of a personality built out of the exploitation of diversity had to wait for Fernand Braudel, at the beginning of the 1980s, but his criticism was devastating (Braudel, 1986).

Geographers have also made a critical contribution to understanding the connections between landscape, memory and identity in their analysis of the ways in which novelists, painters, photographers and other artists used the idea of landscape in their interpretation of national realities. Many of these interpretations utilize landscape as a 'veil' to cover the internal contestations and contradictions of national societies. Denis Cosgrove (1984), for instance, argued that the landscape architecture so popular among Venetian aristocrats in the sixteenth and seventeenth centuries, and British ones in the eighteenth and nineteenth centuries, was intimately linked to their class interests. Beautifying the landscape was a way for them to be accepted by the establishment and legitimate the power they exercized over the lower classes. In the US or Canada, the meaning given to natural landscapes often acted as a way of denying any rights to native populations. Contemporary analysis of the policy of Natural Parks in South Africa discloses similar strategies for negating the role of indigenous populations (Brooks, 2000).

But it would be a mistake to believe that the contribution of geographers' landscape studies to the contemporary debates on identities is only a critical one, as it has also begun to provide those who are conceiving new ways of understanding cultural landscapes with part of their building blocks. One of the most obvious examples is work undertaken recently by Kenneth Olwig, who has developed a very thorough investigation of the origin of the word and idea of landscape in the North Sea area (Olwig, 2002). The term was first used for designating small communities in the coastal region of Schlewsig and Holstein, on the border between Denmark and Germany. As used at that time, a landscape meant at the same time a small area ruled through relatively democratic institutions, the special forms of land organization (the landscape in its modern sense) which characterized it, and the community which achieved a real equilibrium within this setting. When reading Olwig, it is clear that real forms of sustainable growth were at work in these small territorial units. The archetypal form of landscape he describes is a wonderful politico-ecological utopia. The place such a society gave to self-organization and government has much appeal for all those who dream of another, such as a non-capitalist, form of contemporary social life.

Conclusion

This chapter has demonstrated that until the middle of the twentieth century, two forms of identities characterized Western societies: local identities, which were passed down from generation to generation as a form of vernacular culture predominantly through oral means, and national or class identities, which became apparent from the mid-eighteenth century, mainly thanks to history and to a lesser degree, ethnology and geography.

What is clear is that the context within which landscape, memory and identity are debated has changed with particular implications. A particular interpretation of history as a juxtaposition of vernacular memory and national history has been replaced by another one: a new historical approach to vernacular history has displaced vernacular memory, and narratives on the dynamics of cultures or nature have displaced older ones, which focused on national history. This transformation has been paralleled by a similar change in the nature of geography. Instead of dealing only with an objective conceptualization of space, contemporary geography explores the way culture is permeating all the forms of spatial organization. In the field of landscape studies, approaches that focus on the meaning conferred on places by the people who inhabit or visit them substitute for the functional approaches imagined during the twentieth century.

The contemporary crisis of identities is responsible for a renewed interest in landscapes by geographers, and indeed for the spatial turn generally across the social sciences. While landscape studies have always stressed the complex relations which existed between social groups and spatial forms, today they have adopted an even greater role in understanding social consciousness through time.

References

Appleton, J. (1996), *The Experience of Landscape*, Wiley, Chichester.

Berque, A. (1986), *Le Sauvage et l'artifice. Les Japonais devant la nature*, Gallimard, Paris.

Berque, A. (1990), *Médiance. De milieux en paysages*, Reclus, Montpellier.

Braudel, F. (1986), *L'Identité de la France, Vol. 1: Espace et histoire*, Arthaud-Flammarion, Paris.

Brooks, S. (2000), 'Re-reading the Hluhluwe-Umfolozi Game Reserve. Constructions of a "Natural Space"', *Transformation*, 44, pp. 63–78.

Cosgrove, D. (1984), *Social Formation and Symbolic Landscape*, Croom Helm, London.

Cosgrove, D. and Daniels, S. (eds) (1988), *The Iconography of Landscape*, Cambridge University Press, Cambridge.

Dale, P.N., (1986), *The Myth of Japanese Uniqueness*, Croom Helm, London.

Dion, R. (1946), 'La part de la géographie et celle de l'histoire dans l'explication de l'habitat rural du Bassin parisien', *Publications de la Société de Géographie de Lille*, pp. 6–80.

Doi, T. (1986), *The Anatomy of Self. The Individual Versus Society*, Kodansha International, Tokyo.

Freyre, G. (1933), *Casa Grande e Senzala*, Maia and Schmidt, Rio de Janeiro.

Jackson, J.B., (1994), *A Sense of Place, a Sense of Time*, Yale University Press, New Haven.

Krenzlin, A. (1957), 'Blockflur, Langstreitenflur and Gewannflur als Ausdrück agrarischen Wirtschaftsformen in Deutschland', in Juillard, E., de Planhol, X. and Sautter, G. (eds), *Structures agraires et paysages ruraux*, Annales de l'Est, Nancy, pp. 343–352.

Lasserre, F. (1993), 'Paysage, peinture et nationalisme', *Géographie et cultures*, 2, 8, pp. 71–82.

Lunn, H. (1963), *The Swiss and their Mountains. A Study of the Influence of Mountains on Man*, Allen and Unwin, London.

Meitzen, A. (1895), *Siedelung und Agrarwesen der West-germanen und Ost-germanen, der Kelten, Römer, Finnen und Slawen*, Hentz, Berlin, 4 vol.

Nora, P. (1984), 'Entre Mémoire et Histoire', in Nora, P. (ed.), *Les Lieux de mémoire*, vol. 1, *La République*, pp. 17–42.

Olwig, K. (1984), *Nature's Ideological Landscape*, Allen and Unwin, London.

Olwig, K. (2002), *Landscape, Nature and the Body Politic: From Britain's Renaissance to America's New World*, University of Wisconsin Press, Madison.

Vidal de la Blache, P. (1903), *Le Tableau de la Géographie de la France*, Hachette, Paris.

Chapter 7

Valorizing Urban Heritage?
Redevelopment in a Changing City

Niamh M. Moore

Introduction

This chapter, in common with others in this section, explores the commodification of heritage, in this case in the urban landscape of Dublin, Ireland's capital. In recent years, the city has undergone a dramatic transformation as entire areas have been demolished and reconstructed in a bid to rid the city of dereliction and decay and develop a new, vibrant profile on the back of significant tertiary sector development. To support this regeneration, and indeed the economic boom affecting the wider national economy, significant investment has been made in infrastructure, especially with regard to housing and motorway construction. But one thing that has become increasingly contested is the relationship between the dual processes of economic development and heritage protection. The most recent high-profile case has been the long-running attempt through the legal system to prevent the destruction of Carrickmines castle, a medieval fortification on the route of the M50, Dublin's orbital motorway. The case concluded unsuccessfully from the perspective of those interest groups broadly representing a heritage-environmental agenda and the motorway opened in late June 2005. But beyond the immediate implications of the decision, this long drawn-out saga has opened a wider debate in Ireland focused on 'how and for what purpose we measure the value of heritage' (O'Keeffe, 2005, 142).

This is continuing through ongoing attempts by objectors to block the planned development of the M3 motorway from Dublin to the northwest of Ireland in a move that would skirt the national monument of Tara, County Meath.[1] This debate follows years of contradictory messages being expressed by those with responsibility for the heritage sector, exemplified in the National Monuments Act, 1930, which was in force until relatively recently and contained a clause detailing the mechanism through which national monuments could actually be destroyed!

What is particularly notable about the most recent debate surrounding the Hill of Tara/Skryne Valley campaign is the level of media sophistication in evidence,

---

1   Embraced as the spiritual capital of Ireland, Tara was the Seat of the High Kings of Ireland.

with objectors harnessing the support of Hollywood actor, Stuart Townsend, and the internet to globalize the debate. The sense that Tara is of European significance and that its history represents an important defining moment in Irish identity is being vigorously played upon. The site is being imbued with an iconic status in an attempt to halt the new motorway, while the campaign is being driven for the most part by those who do not need the new infrastructure and can afford to romanticize the image of Tara. The fact that the new motorway will be further away from the heritage site than the existing road is being completely ignored. Objectors do not accept that for those residents of County Meath, dependent on swift access to Dublin for work and other functions, Tara, while a location to be 'preserved and understood [...] [must] also be modified to meet the needs of a changing world' (Lowenthal, 1985, 239).

Yet it is not only at this national or regional scale that the debate surrounding Ireland's heritages has gathered pace and intensity. At an intra-urban scale over the past thirty years, a number of sites throughout the city of Dublin have been demolished for redevelopment, none more controversial than the development of the Civic Offices over many parts of the medieval town at Wood Quay. Even in the late 1970s, cultural politics was quite evidently at work in the comments of Senator Gemma Hussey during the parliamentary debates:

> I would like to remind the Government at this point – this Government which makes such an extremely strong case for the preservation of the Irish language – that they should remember that the Wood Quay question seems to many people to be of extreme importance, no less than making a big effort to preserve our language. (Seanad Éireann Parliamentary Debates, 6 December 1978)

The issue of heritage-politics therefore is nothing new in an Irish context, but it is the nature of this discourse that has begun to change in tandem with the transformation of Irish society in recent decades. Although preservation of the built heritage continues to be a central theme, there is an emerging debate surrounding the consumption and private appropriation of public heritage, particularly in areas subject to urban renewal and revitalization. This is a central theme of this chapter, which examines the redevelopment of Stack A in the north docklands from the perspective of cultural heritage. The motives and impact of the redevelopment and marketing choices taken at this site are related to broader discussions on the value and meaning of heritage in a post-industrial city.

Heritage, Memory and the City

In considering the discourses surrounding the Carrickmines–M50 motorway controversy, O'Keeffe argues that one of the reasons why the general public failed to be mobilized and exercized about this particular issue was that 'the public has found little at Carrickmines at which to "remember" its past or in which to invest any part of its identity' (O'Keeffe, 2005, 148). Although personal identity is perhaps considered relatively static, place identity is more dynamic, particularly in cities built upon innovation and dependent upon change for survival. Massey (1992, 11) describes places as being 'formed out of the particular set of social relations which interact at a particular location. And the singularity of any individual place is formed in part out of the specificity of the interactions which occur at that location'. As the context changes, so too will the character of the area. However, cities are not simply blank canvases on which new stories can be written and although undergoing change, do remain intimately tied to their past. The landscape of a particular place will tell us much about the history of the people, particularly if the memory held by those living there or associated with it the longest remains strong. In a world characterized as runaway and constantly in flux, memory is critical in the formation of both personal and place identity but it is also crucial in shaping discourses on preservation, development, and how heritage is defined and represented (Giddens, 2002).

In order to form identities, Lowenthal (1985) asserts a dependence on three key attributes: relics, history and memory. But it is the contingent nature of the latter that directs in so many ways our reactions to the former and results in the contestation of place between different interests. Partly this is because 'memory transforms the experienced past into what we later think it should have been, eliminating undesired scenes and making favoured ones suitable' (Lowenthal, 1985, 206). This element of 'forgetting', however significant, is largely subconscious, but yet it plays a crucial role in the production of both personal and place memory and identity. As Winchester has remarked 'each person or group views, uses and constructs the same landscape in different ways; these are neither "right" or "wrong", but rather are part of the many layers of meaning within one landscape' (Winchester, 1992, 140). It is therefore not difficult to understand how landscapes that are particularly unique, legible and imageable within a particular city, will become sites of contestation during economic and physical restructuring.

At a global scale, economic restructuring and physical change have gone hand-in-hand and have recently led to debates on the nature of the city and the kinds of development that we should be promoting. In response to a decline in the 1960s and 1970s, associated with global economic change and the new international division of labour, many national and city governments were faced with derelict and degraded city centres, high unemployment and negative public perception. Responding to this and building on a successful development model that emerged from Boston, city managers recognized the necessity to find alternative ways of generating wealth, reasserting the position of their cities, and demolishing the negative perceptions that abounded (Ward, 1998). A more 'entrepreneurial' approach was adopted to attract investment and economic development, through the active repackaging and remoulding of cities. Central to this new approach to urban governance has been the marketing of place, resulting in the inclusion of professional marketeers and outside consultants in the revitalisation process. City promotion has become an intensive PR exercise, with the evolution of the 'right image' to attract 'the right sort of people' of paramount importance. It could easily be argued that understanding city-image building or 're-imaging' is the key to comprehending the dynamics of city development and management in the 21st century, as the city is now, like any other commodity, being marketed and sold.

Although Ward (1998, 236, 240) argues that 'episodes of place selling are mainly associated with periods of economic change and with urban systems or parts of urban systems which are experiencing that change', he also indicates that 'typical images still exclude much that makes up the reality of place or they appropriate aspects of place in ways that narrow meaning'. The aim of the new 'urban entrepreneurs' through public-private partnerships and other strategies is to neutralize the negative and elevate the positive images of a particular place through pro-active marketing strategies. So, while urban professionals actively engage in constructing place, the relics and indeed local memory may end up significantly out of place. Disagreements arise between the so-called 'objective history' of an area and the memory of the past held by individuals and collectively by communities. But although this kind of memory, in terms of its construction and content, has already been accepted as subconsciously selective, it embodies a passive sense of forgetting.

Place marketers on the other hand pro-actively subject the past to a very controlled and deliberate reinterpretation. Images selected for promotional purposes actively and purposely exaggerate the positive aspects of urban living, while glossing over unfavourable urban realities in an attempt to promote the ideal post-industrial city, replete with museums, art galleries, theatres, festival retailing and high-income housing. Winchester has remarked that 'this marketing of place often constitutes a major break with the industrial past, as themes and events are used to manufacture a new present. In the reconstruction of place, some histories are privileged while others are expunged from the collective memory... In the process of place marketing, history and landscape are commodified into saleable chunks' (Winchester, 2003, 135).

This strategy has not been without its critics who argue that the new entrepreneurial approach adopted in city governance has marketed the city selectively towards a small, wealthy elite and disenfranchized those for whom the memory of a place is often strongest. Those who have occupied a place for longest are, after the work of the place marketers, often those who feel most 'out of place'. As well as this social dimension, the economics of place marketing are often not as advantageous as assumed given that many aggressive strategies have been nullified by replication in a period of intensive inter-urban competition (Boyle and Hughes, 1991). In recent years, this has been compounded by the addition of a new dimension, increasingly intense intra-urban competition. Within cities, each quarter and developer is now competing with others for investment as certain districts receive favourable incentives for regeneration. This has been nowhere more apparent than in the city of Dublin where development is now so widespread and competition so fierce that the construction industry and promoters are looking for new ways to obtain 'an edge' and sell their development. Increasingly 'heritage' is being appropriated as the key to doing this.

The Irish Context

Until relatively recently, heritage in Ireland has been conceived in a very narrow sense both temporally and spatially, due in part to the lack of legislative and other policy documents drawing attention to the subject. In a discussion on heritage in Ireland, the Heritage Officer for An Taisce, Ireland's national conservation body, argued that 'we now regard the legacy of pre-historic, early Christian, medieval and Georgian Ireland as our heritage and it is the image in which this country is marketed' (*Irish Independent*, 2 November 2002). Until the amendment of the National Monuments Act in the last decade, Irish heritage was deeply associated with antiquity and a pre-modern world and promoted in ways that portrayed a certain sense of Irishness, one that was predominantly rural, male and Catholic (O'Connor, 1993). In the early decades of the twentieth century, heritage was enmeshed in a nationalist agenda and an active attempt to construct a particular kind of Irish identity. Although this is widely recognized by many authors, McCarthy (2005, 6) suggests that in the last decade, perhaps driven by the changing socio-political and cultural position of Northern Ireland, the adoption of more pluralist approaches to culture and identity 'have served to successfully rescue many forgotten historical events from the dustheap of Ireland's histories'. In the contemporary world where the battle lines have moved away from a rather narrow definition of 'heritage', he suggests that an even greater challenge comes from the conflicts arising from the dichotomy between 'heritage as a cultural resource and heritage as a capitalist item for consumption' (McCarthy, 2005, 10). Although he suggests that commodification is, above all, driven by the heritage industry, it is worthwhile considering the other agents and sectors involved in this activity.

In general, tourism has been closely associated with this process and McManus discusses in detail the efforts that have been made over the last twenty years in Ireland to create a new heritage in urban areas for the benefit of tourists (McManus, 2005). She believes that this progression has occurred due to broader changes in socio-economic and demographic trends in Ireland that have seen the country become highly urbanized since the mid-1990s. Portraying the country's heritage and lifestyle as predominantly rural and traditional was totally at odds with the reality of twenty-first century Ireland, yet it perpetuated the notion that 'urban centres were not … truly Irish but were seen as "foreign" imports' (McManus, 2005, 237). However, even in this shift towards a more urban-based definition of Irishness, significant gaps remain. An appropriate illustration comes from the Dublin City Heritage Plan, which states:

> When we consider the heritage of Dublin City some things immediately spring to mind, Viking archaeology, Georgian architecture, diverse wildlife habitats. We all have different perceptions as to the definition, value and future of Dublin City's heritage. This Plan is an attempt to harness all of these diverse and often contradictory opinions into a co-ordinated vision of the future for our capital's heritage. (Dublin City Heritage Plan, 2002, 6)

Although aspirational and worthy, this introduction highlights one of the critical issues in heritage promotion, management and urban development at present – how to deal with nineteenth-century memory, history and relics? Although the definition

Figure 7.1   Stack A Prior to Redevelopment
Source: Niamh Moore

of heritage has now been broadened to include the Viking, medieval and Georgian urban heritage, what place do the relics of the industrial era – the antithesis of all that is traditionally Irish – have in our mindsets and places? And this question gains further salience when we consider how so many former industrial districts are now subject to policies of renewal and revitalization that are re-writing their physical and social fabric. In the remainder of this chapter, one icon of Dublin's, albeit modest, industrial maritime era are considered – the Stack A complex in the Docklands area of the north-east inner city. Drawing on the suggestions of McCarthy (2005, 36) that a 'debate on the commodification of the past is vital, especially if one is to appreciate how such usage impinges upon constructs of memory and identity' the chapter concludes with a discussion on the relationship between Dubliners and their nineteenth-century maritime history.

Stack A: A Jewel in the Crown

From the beginning of discussions in the mid-1980s on urban redevelopment in Dublin, the Custom House Docks project was portrayed as the flagship scheme to highlight the benefits of a more entrepreneurial approach to revitalization. Central to the Master Plan drawn up by the development agency, the Custom House Docks Development Authority (CHDDA), was the building within the complex known as Stack A (Figure 7.1).

A former bonded tobacco warehouse, built of brick with one of the finest iron roofs in Europe, Stack A was described by G.N. Wright in 1821 (the year of its opening) as an 'ingenious construction' (Wright, 1821, 7) and since the 1980s it has been afforded protected structure status (Figure 7.2). But beyond its architectural significance, the

Figure 7.2   Interior Structure of Stack A Prior to Redevelopment
Source: Niamh Moore

building is historically important given that it was the only building in mid-nineteenth Dublin sufficiently large to hold a banquet in honour of the Irish Crimean War veterans. Murphy (2002), in his treatise on Ireland's role and reaction to this campaign, suggests that the Irish public followed developments with great interest, publicly supporting the troops leaving Ireland and the tone of ballads sung suggest that Irish opinion was in favour of the war. This substantial banquet was widely reported in the contemporary newspapers and there is little doubt that Stack A must have been familiar to many Dubliners. The contemporaneous *Freeman's Journal*, for example, reported that:

> There were laid 250 hams, 230 legs of mutton, 250 pieces of beef, 500 meat pies, 100 venison pasties, 100 rice puddings, 250 plum puddings weighing one ton and a half, 200 turkeys and 200 geese, 2,000 rolls, 2,500 lbs of bread, 3 tons of potatoes, 8,500 quart bottles and 3,500 pint bottles of port. (*Freeman's Journal*, 23 October 1856)

Mindful of its architectural and historical significance, and the heritage potential of the building, the original plans for renewal in 1987 proposed a range of uses for Stack A. These included a night-club/bar area, winter garden, and a cultural/museum attraction to promote vibrancy but also to ensure that the historic fabric would be conserved and open to the general public. This was reiterated in the Master Project Agreement signed between the developers and the CHDDA, which suggested that Stack A would become a hub of activity in docklands, setting itself aside from developments in the rest of the city through the exploitation of its maritime history and environment. Yet at the time of writing, eighteen years on from the original plans, ordinary Dubliners have still not, for the most part, had the opportunity to enjoy and marvel at this remarkable testament to nineteenth-century innovation and engineering. Since the original proposals, the plans for Stack A have been reconstituted numerous times and this historic building, although perhaps providing the greatest potential for revitalization has proved to be the most difficult element of the docklands project to deliver.

One of the greatest delays to conservation work on the building was that the Hardwicke/British Land consortium of developers, who had exclusive rights to the building in the late 1980s and early 1990s, favoured the development of a shopping centre in the historic warehouse and attempted to have the facility re-zoned to retail use. After much debate, this proposal was rejected and the CHDDA conceded on 1 December 1994, that it 'has a legal obligation to provide a genuine museum in the complex'. What this would be has remained highly contentious in the last ten years, not because of lack of interest but for other reasons. In fact since the original Master Plan was published a range of agencies and other groups have suggested a myriad of potential uses for Stack A, all with a cultural and public access dimension (Table 7.1). Each of them were rejected on financial or operational grounds, and a final call for proposals was made by the Dublin Docklands Development Authority (DDDA) in March 2001.[2]

However, it is very likely at this stage that the DDDA had their own clear vision for the building, given that they significantly reduced the space being made available to cultural/heritage uses in favour of more commercial options. This was highlighted in an internal departmental briefing note to the Minister of Arts, Heritage, Gaeltacht and the Islands in July 2001 which stated that:

> the space on offer from DDDA in Stack A has ranged in their correspondence from a low of 36,900 square feet for the cultural facility to a high of 48,000 square feet … If the final concept of the museum put forward by us does not appeal to their commercial sense, they will oppose it.

By February 2002, the DDDA wrote to the proponents of all the proposed projects stating their preference to work with the Department of Arts, Heritage, Gaeltacht and the Islands on a proposed Museum of Dublin. The proposal originated in 1999 when the Minister for the Environment, Noel Dempsey, had suggested the development of a social history and transport museum at the site. This proposal was further refined by the Department of Arts, Sport and Tourism and a decision was made to focus the project on a much narrower theme, *Dublin: Its History as a City and Port*. One of the reasons that this idea appealed to the Minister was the fact that Dublin City Council had already established a Working Commission for the city of Dublin Museum and had progressed discussions with the Docklands Authority. The Stack A building was an ideal location for such a facility being located in the interstitial zone between the contemporary city and port, of historic significance itself and standing as a monument to the great maritime tradition that once existed in Dublin.

---

2   The CHDDA was disbanded in 1997 and replaced by the DDDA, to redevelop a much larger area of approximately 1300 acres north and south of the river Liffey.

Table 7.1   Proposed Uses for Stack A, 1986–2001

Date	Proponent	Project
April 1986	Dept of Environment	Commissioned study: Museum of Modern Art
May 1986	Dr Danny O'Hare, DCU	Science, Technology, Children's Museum
Oct 1986	The Arts Council	Gallery of Modern Art
April 1987	Dr Mollan, RDS	Children's Museum
June 1987	National Museum Ireland	National Folklife Museum
Oct 1987	Dept of An Taoiseach	Gallery of Modern Art, Folk and Science Museum
Nov 1987	Dept of An Taoiseach	Museum of Decorative Arts & Folklife
Jan 1988	RDS	Science Centre
May 1988	Department of Marine	Maritime Museum
June 1988	Irish Museums Trust	Creative Participatory Centre for Young People
Feb 1989	An Post	Irish Postal Museum
May 1989	National Museum Ireland	Museum of Postal History
Nov 1989	National Committee for Science	History of Science & Technology Museum
Jan 1990	Royal Irish Academy	History of Science & Technology Museum
Feb 1992	Stan Nielsen, EOLAS	Science Centre
Oct 1997	Irish Museum Modern Art	Second site for the IMMA
Mar 1998	Dorothy Walker	Proposals for a Scully Museum (Sean Scully)
Mar 1998	Maritime Institute of Ireland	Maritime Museum
Nov 1998	Report by CHDDA	Transport/Social History/Children's Museum
Apr 1999	Scroope Design	Proposal for Explorarium
Sept 1999	Dept of Environment	Social History oriented Transport Museum
Sept 2000	Chester Beatty Library	Museum of City of Dublin
1993–2000	DISCovery	Science Museum
Dec 2000	National Museum Ireland	Museum of Social History and Transport
Dec 2001	Dept Arts, Sport and Tourism	Museum of Dublin: A City and Port

Source: Compiled from Department of Arts, Sport and Tourism

The initial submission from the government department to the development authority requested that the allocated floor area would need to be increased to 60,000 square feet to guarantee the viability of the project, while the amenity should also have operational independence and total autonomy from the DDDA. The rationale given by Professor Loughlin Kealy of UCD, as per the minutes of the meeting, was that he had 'reservations about the concept of the museum being pushed into a space left over from something else, i.e. the various retail outlets and commercial ventures envisaged by the DDDA'. At the same meeting, Philip Maguire, Assistant City Manager concurred and explained that while the DDDA 'had a preference for a cultural usage … commercial operation was at the core of their activities'. A letter from the DDDA of 11 November 2001 in response to these queries reinforced this viewpoint stating that there was no way these requests could be granted and 'that there is no possibility of it extending to 60,000 square feet or to taking over the total ground floor area of Stack A'. Given the inability to plot a way forward, the government decided in December 2001 to authorize a feasibility study in order to gauge how the project might be taken a step further. Meanwhile a sense of urgency seemed to be growing as Peter Coyne, CEO of the DDDA, warned the Department on 21 January 2002 that unless the feasibility study was undertaken rapidly 'the opportunity to consider Stack A as the home for the Museum of Dublin may be lost'. At the same time as the Government was being forced to fast-track the study and DDDA were warning that the future of the project could hang in the balance, they were also writing to other museum candidates turning down their proposals and indicating that the Museum of Dublin was almost a certainty. The inordinate power of the DDDA and the pressure that it appears to have been able to bring to bear on Government is extraordinary, given that it is in statutory terms answerable to and controlled by the Minister for the Environment, and by extension the Government.

The report commissioned by the Government concluded that the project was 'likely to be feasible, practical and sustainable as well as making a valuable contribution to Dublin's self-esteem, cultural life and tourism product', but the consultants appointed by the Authority, Locum Destination/At Large, concluded contradictorily that visitor numbers were unlikely to be sustainable and that they would recommend a contemporary art facility for the building. Further disagreements ensued and in August 2002, the DDDA Board indicated that all negotiations with the Department would be concluded. In December 2002, the Government made a final decision to 'withdraw from further involvement with Stack A and leave the DDDA free to pursue the type of cultural facility that meets their requirement'.

The type of cultural facility that did meet the requirements of the DDDA was a small events and high-end retailing venue and a €50 million building conservation project was undertaken. Now known and marketed as 'chq', Stack A is set to become a luxury shopping quarter, with five restaurants and 15,111 square feet of event and exhibition space. Terence Conran may be the major restaurant anchor, Cabot & Company will operate a wine bar and private wine club, while other major anchors are being sought to replace the hoped-for anchor tenant of Harvey Nichols. Doubts have been raised as to the viability of this kind of development at this location, and comparisons have been made with London's failed Tobacco Dock development in Wapping. Coincidentally, the proposals for Stack A designed by the same architect and built six years after its London equivalent, are very similar. Part of the original plans for a shopping and entertainment centre in London Docklands, the Tobacco Dock warehouse went into decline very soon after opening and is now subject to renewed plans to regenerate the area.

In Dublin, conservation work has now been completed and a new glass frontage on to the Liffey and glass wings facing on to George's Dock have been added to the façade of Stack A (Figure 7.3). The result has been a physical reopening of this

Figure 7.3   Stack A after Conservation and Redevelopment
Source: Niamh Moore

part of the Custom House Docks site to the river and the city, but although due to open fully for business in summer 2004, chq has still not been completely let and is only partially opened. However a greater concern is that given the marketing of the project, the facilities provided and the kind of clientele it is aiming to attract, this amenity will lose its potential to entertain and inform the general public about a much-forgotten and ignored part of the heritage of Dublin.

Conclusion: Reflecting on the Past

The difficulties surrounding the redevelopment of one of the key relics of nineteenth-century Dublin, Stack A, exemplifies the contention that the 'built heritage under various renewal schemes has been seen as a commodity valued in economic and tourism terms rather than for its cultural or social significance' (McManus, 2005, 242). The prolonged debate, over more than twenty years, regarding the future of this building highlighted the reluctance on the part of the development authorities to engage in the conservation of a building and promote it for heritage reasons. Only when an economically profitable vision of its future became dominant did the DDDA invest in conservation work. This is not unusual as often these broader societal frameworks drive our definition and attitude towards heritage. The way in which popular memory is amalgamated into the new stories being woven depends on the salience and strength of the original place history. Usually the potential of a place to tell us something about the lives of ordinary individuals in the past is given less credence than those places that represent the extraordinary. So while we revel in the history and elegant lifestyles of the elite through visits to places like the National Museum of Decorative Arts, the old parliament at the Bank of Ireland on College Green, Powerscourt House and other monumental buildings throughout the city, where do we hear or see the stories of ordinary Dubliners, who witnessed some of the most dramatic events in Irish history from their tenement homes?

In her discussions of redevelopment at Battery Park City (New York), Boyer (1996) draws on a motif utilized by many other authors in their expositions of revitalization at former maritime districts such as the Victoria and Albert Waterfront at Cape Town or South Street Seaport, New York. As place and place history is promoted or marketed to newcomers and outsiders, it is radically sanitized. Place marketing and imaging in Glasgow, associated with the attraction of investment and tourism, has also been criticized for burying 'the facts of the past which have become inconvenient for its new, sanitized, and marketable image of the city' (Donnelly, 1990). In Dublin Docklands, this disjuncture between the marketing machine and the memory of local residents is even more apparent. Community groups argue that their memory and identity is being devalued as the construction of a new place identity, based on the prominent position of the Dublin Docklands in international finance networks, fails to recognize the less-polished, 'everyday' history and diminishes the significance of the relics of the maritime heyday. The North Wall Community Association has argued that:

> There is nothing to reflect the culture or history of the area. Sadly Stack A, which was ring-fenced back in 1992 for that specific purpose is to become an area of bars and restaurants. When you look at the Albert Docks in Liverpool with its Maritime Museum, reflecting the great seafaring history and tradition of Liverpool [...] you can get a sense of what could have been achieved in Stack A. Instead very shortly the developer driven concrete jungle will be complete and the sands of time will wash away the memory of the history of the Dockland [...] who will recall the Guinness boats tied up alongside Custom House Quay? [...] Who will recall the men who worked on the docks, the men who dug the coal boats with their no. 7 shovels? [...] The lives of the ordinary people and their daily struggle to survive [...] who will remember? Who will tell their story? (*North Wall Community Newsletter*, Easter 2005)

Instead Stack A has been marketed in a highly selective manner and rather than providing a place for people to congregate, chq may become a new symbol of segregation and exclusivity in an already divided district. The development authority propose that:

> CHQ will host exhibitions and offer a dynamic and diverse programme of events. This will include exhibitions of contemporary interior design, visual art, photography, sculpture, fashion and will be complemented by an evolving schedule of tenant showcases, high value product and promotional launches. Lectures and seminars, corporate sponsored events, seasonal themed events and more [...] Importantly chq event management will retain control of programming and ensure that all events staged at chq are entirely supportive of the chq brand values and of the individual commercial tenant brands. (chq marketing brochure, 2005)

So often 'heritage is reduced to little more than an adjunct to urban tourism and place marketing' and the redevelopment of Stack A appears to be yet another example (Graham, 2002, 1013). The exclusive nature of this development, and its incongruence with the history of the area and the building itself, was foreseen in a report to Government by Event Ireland Ltd. They declared that such a combination 'will not assist a policy of social inclusion if the retailing environment appears hostile to lower income groups', and contrasted it with the Museum of Dublin that aimed to 'have a socially-inclusive marketing policy and will not attempt to confine itself to an upmarket audience' (Event Ireland, 2002, 35). The intersection of state power and private capital at this site has produced a landscape that is both the outcome, and the medium, through which new power politics are being played out. To return to the comments made by Senator Gemma Hussey during the Wood Quay parliamentary debates in December 1978:

> Just because the whole business of Wood Quay has become a *cause célèbre* it should not be allowed to be made into any kind of a political football or made into a matter of somebody's pride or bad feeling or the subject of some kind of gamesmanship. (Seanad Éireann Parliamentary Debates, 6 December 1978)

Similar comments could be made regarding the redevelopment of Stack A. It would appear that a massive opportunity has been lost to embrace the maritime and industrial past of Dublin as a city and port. The lack of any firm government control on development at this site has resulted in a zero-sum game from both an economic and heritage perspective. Although a dichotomy did emerge in the last decade between those who wished to see the facility conserved for public access and cultural use and those who preferred a more commercial option, the bottom line is that the part of the building that is open is very specifically targeted.

Although the history of Stack A does not fit with the traditional concepts of Irishness, or even more recent definitions of Irish heritage which have included some urban aspects, the facility did have the potential to play a major role in contributing to further debates on Dublin's history and identity. Instead the state has legitimized its lack of action in becoming a custodian of this heritage building through arguments surrounding the need to support entrepreneurialism and private sector activity. The ability of a state-established development agency to brow-beat a government into withdrawing a proposal for a heritage facility in such an important location is incredible and demonstrates a reluctance on the part of the legislature to take heritage seriously. The result has been a displacement of identity for local communities, as their memories and the goals of contemporary place marketing are diametrically opposed. At a broader scale, the continuing lack of activity at the Stack A site, combined with the privatization of other significant relics of the industrial period such as the Clayton gasometer in the south docklands has culminated in the continued marginalization of nineteenth-century urban relics as a part of Ireland's, and more specifically, Dublin's heritage. It really would appear that, 'In talking about the past, we lie with every breath we draw' (Maxwell, 1980, 29).

References

Boyer, M.C. (1996), *The City of Collective Memory: Its Historical Imagery and Architectural Entertainments*, Cambridge University Press, Cambridge.

Boyle, M. and Hughes, G. (1991), 'The Politics of the Representation of "The Real": Discourses from the Left on Glasgow's Role as European City of Culture, 1990', *Area*, 23, 3, pp. 217–228.

Donnelly, M. (1990), 'The Dirty Immoral Witch Hunt that is the Elspeth King Affair', *Glasgow Herald*, 29 August 1990.

Dublin City Council (2002), *Dublin City Heritage Plan.*

Event Ireland (2002), *Feasibility Study for the Redevelopment of Stack A*, Event Ireland, Dublin.

Giddens, A. (2002), *Runaway World: How Globalization is Reshaping Our Lives*, Routledge, London.

Graham, B. (2002), 'Heritage as Knowledge: Capital or Culture?', *Urban Studies*, 39, 5/6, pp. 1003–1017.

Lowenthal, D. (1985), *The Past is a Foreign Country*, Cambridge University Press, Cambridge.

Massey, D. (1992), 'A Place Called Home?', *New Formations*, 7, pp 3–15.

Maxwell, W. (1980), *So Long, See You Tomorrow*, Ballantine, New York.

McCarthy, M. (ed.) (2005), *Ireland's Heritages: Critical Perspectives on Memory and Identity*, Ashgate, Aldershot.

McManus, R. (2005), 'Identity Crisis? Heritage Construction, Tourism and Place Marketing in Ireland', in McCarthy, M. (ed.) *Ireland's Heritages*, Ashgate, Aldershot, pp. 235–250.

Murphy, D. (2002), *Ireland and the Crimean War*, Four Courts, Dublin.

O'Connor, B. (1993), 'Myths and Mirrors: Tourist Images and National Identity', in O'Connor, B. and Cronin, M. (eds) *Tourism in Ireland: A Critical Analysis*, Cork University Press, Cork, pp. 68–85.

O'Keeffe, T. (2005), 'Heritage, Rhetoric, Identity: Critical Reflections on the Carrickmines Castle Controversy', in McCarthy, M. (ed) *Ireland's Heritages*, Ashgate, Aldershot, pp. 139–51.

Ward, S. (1998), *Selling Places: The Marketing and Promotion of Towns and Cities, 1850–2000*, Spon, London.

Winchester, H.P.M. (1992), 'The Construction and Deconstruction of Women's Roles in the Urban Landscape', in Anderson, K. and Gale, F. (eds) *Inventing Places: Studies in Cultural Geography*, Longman, Melbourne, pp. 139–56.

Winchester, H.P.M., Kong, L. and Dunn, K. (2003), *Landscapes: Ways of Imagining the World*, Pearson, London.

Wright, G.N. (1821), *An Historical Guide to Ancient and Modern Dublin: Illustrated by Engravings, after Drawings by George Petrie, Esq. To which is Annexed a Plan of the City*, Baldwin, Cradock and Joy, London.

Chapter 8

Moving Buildings and Changing History

Stephen F. Mills

Introduction

Moving buildings is rarely considered problematic. And anyone old enough to remember the film *Deliverance* will recall that moving buildings out of the way of flooding has long been a technically routine matter, however disconcerting for those whose lives are so disrupted. Back in the 1950s, Pathé News showed a home trundling along an Ontario road out of the way of the St Lawrence Seaway construction. But why would buildings ever be removed other than to make way for redevelopment? And when they are removed, so what? Can buildings in their new locations ever retain any of the nuances of time and place, which were embedded *in situ*, or do such buildings become elements in the ever greater 'placelessness' that many argue is so much part of the post-modern condition (Relph, 1976)?

Buildings have long been copied rather than moved. In the Middle Ages St Sepulchre in Northampton was built to recall its Jerusalem prototype. Many other buildings that might seem to have been relocated, like the National Cathedral in Washington DC, are actually relatively new, the 'medieval' west window having only been completed in the 1970s. Other buildings, mainly iconic ones, have been copied as part of business and entertainment complexes. Theme parks, and now resorts such as Las Vegas, are awash with reconstructed Pyramids, Eiffel Towers, Leaning Towers of Pisa, and Venetian plazas. If one names a US town Athens, then an Acropolis cannot be far behind. But there are other buildings that have appeared in the New World that are far from being either new or recycled. Many are authentic, having been taken down and transported across the ocean. Can such authenticated buildings challenge the tide of replicas that increasingly promote placeless simulacra over so much of the modern world?

Moving artefacts large or small didn't start with the North Atlantic passage, however. But there are few, if any records before the modern era, of the deconstruction and subsequent reconstruction elsewhere of a building. Of course Vikings took their expertise to the land of the Inuit, and replicas of their minimalist structures can be seen today at L'Ainse Aux Meadows in Newfoundland. Later settlers took over building technologies ranging from log cabins from Finland to Commonwealth era church architecture from England (Jordan and Kaups, 1989). Today the Americas are covered with buildings, religious and secular, that recall France, England, Spain, and even, as at Fort Foss north of San Francisco, Mother Russia. But when these were constructed they were new and not relocated from earlier old world sites. Designs, rituals and the word of the Lord were sufficient to provide each new site with the requisite pedigree. Any physical structure larger than a tent was just too immutable, unlike the memories of such buildings taken far and wide. Buildings were thus either ephemeral and not worth taking down and reconstructing, or semi-permanent fixtures in the landscape that could only be ignored or at best destroyed. But today buildings have been taken down and relocated often thousands of miles away. What does this say about our modern connections with the past and with other places? What links do such placeless structures provide with the times and places from which they have sprung?

Moving Buildings

Moving buildings is still very much a minority sport, with few groups and fewer individuals prepared to take the requisite time and effort. It seems to have been in the nineteenth century that buildings were first moved in any systematic way. As the world became smaller with the arrival of the railways, the telegraph, and the rise of a new imperial order, certain buildings started to take on novel roles that both challenged and sustained this new order. And if certain societies were special, then the buildings that had emerged from their passage to this place were presumably also special and worthy of preservation. Thus buildings began to take on new meaning and were moved, many into the new genre of the folk museum, lauding that which was already fading from people's daily experiences.

The rise of the folk museum quickly came to involve not just the acquisition and presentation of country crafts and artefacts, but the provision of suitable buildings within which such items could be displayed. And what could be better in such circumstances than buildings that were themselves exhibits? Open-air museums in Europe began to collect buildings as well as interiors. From Stockholm to Copenhagen, Århus, Odense, Lillehammer and Reykjavik open-air folk museums spread across Scandinavia, and are now found across Northern Europe, particularly in Germany, Belgium, Switzerland, Britain and Ireland. These are far from antiquarian collections of curiosities as most, if not all, have been collected for very specific reasons. They illustrate and some would say even help create a sense of belonging, if only through a public familiarity with the historically-specific degree of cultural variety permitted within the emerging state. Collections have thus been used didactically to portray a sense of who we are, and by inference, to distinguish ourselves from those with different customs, folklore and vernacular architecture.

Buildings in this genre were removed and preserved to represent a lost past for future generations. But this strategy was not limited to European folk movements and their supporters in the museum world. Imperial expositions were equally keen to present the past through the presentation of dioramas recreating times before, during and after the advent of 'civilisation' (Mills, 1996). Sometimes this involved, for example, the building of Japanese structures during the very exposition itself, at other times merely the presence of primitive African villagers in the shadow of the modern industrial Eiffel Tower. Such expositions, though, generally involved the construction of replicas. Their varying degrees of authenticity ranged from little more than canvas stage props, to carefully recreated buildings, true in every detail (Greenhalgh, 1988). This tradition can still be seen in the carefully replicated buildings such as the Norwegian stave church at Disney's Epcot, or the museum equivalent at Moesgård Manor near Århus in Jutland.

Buildings are of course only very large artefacts, and the relocation of artefacts has long been recognized as problematic. To address the dislocation of artefact from its surroundings, small items have increasingly been presented within not so much an array of similar artefacts as at the famous Pitt Rivers collection, but instead have been associated with related items, as in kitchen utensils being displayed with other kitchen utensils in a kitchen mock up. Some artefacts are so large, such as the Elgin Marbles, that they have had to be displayed within specifically constructed rooms, large enough to allow the artefacts to be displayed with regard to their original geometry. In adjacent rooms small temples have been reconstructed from around the Mediterranean and Asia Minor. But being within larger buildings thousands of miles from home such artefacts are clearly removed from their original contexts, for good or ill. Being within the British Museum, the Louvre, Berlin's Altes Museum or the Smithsonian, such artefacts, large or small, are clearly just that, exhibition pieces that are no longer *in situ*. No matter how many theatrical devices are used to remind the viewer that such artefacts were once embedded within wider arrays and historically contingent cultures, no one can mistake Bloomsbury for the Mediterranean, or really believe that African huts along the Seine are Africa. But the more recent relocation of buildings to rural open-air museums results in something else: an attempt to blur the distinction between museum exhibit and heritage site (Mills, 2003).

Many historic buildings have been deliberately destroyed as part of wartime action: Pushkin's house was deliberately destroyed during Operation Barbarosa. Others have been destroyed because of their wider symbolic role; witness the destruction of places of worship across so much of the Balkans, the dynamiting of mosques and the torching of churches, similar to the Kristallnacht fate of synagogues across Germany. But what of buildings that were removed? Few armies have been able to take their buildings with them, and fewer refugees have been able to flee with anything more than memories from which they might later come to recreate if not restore their precious surroundings. Few apart from cultural ethnographers have sought to move buildings intact (though armies are notorious for stripping buildings of their fittings and fixtures). With the riches of the New World some American entrepreneurs have however sought to buy not just furniture and effects from the Old World, but have sought whole staircases, chimneys, porches, windows and so forth, creating pastiche structures out of what they had shipped in, as in the famous summer 'cottages' of Rhode Island artefacts, assembled in ways the individual items had never known before. But in a few cases millionaires have sought to bring back complete buildings, like those along the James River in, and around, Richmond

(Virginia). Tobacco millionaires purchased, took apart, shipped and recreated several buildings that they desired to live in back home, though not always with the original furniture and effects, which were usually far more eclectic than the buildings' original inhabitants would have enjoyed. Why they did this can only be inferred, but the building of houses with instant pedigree, in a State so besotted with historical images and structures outside the museum world, must surely have had something to do with the needs of the *nouveaux riches* to assert some status to which they felt they were entitled.

One excellent example is Agecroft Hall, a fifteenth century, half-timbered manor house brought over from Lancashire to be assembled in the late 1920s by T.C. Williams Jr, heir to a tobacco fortune. Filled with a collection of paintings, armour, musical instruments, tapestries and furniture from the Tudor and early Stuart periods it stands within 23 acres presented in full period style. Next-door is Virginia House, another Tudor mansion brought over at about the same time. This house has a more eclectic array of English and Spanish pieces, many from the sixteenth century. Authenticity was not here the overriding concern unlike the English farmstead at the Staunton museum.

A New Role for Heritage Buildings

But now there is a new rationale for moving buildings: conservation and education. This is particularly so where the buildings are brought great distances and are aimed not just to confirm a local identity to local visitors, but to make some statement about long distance transfers of culture and identity in the recent past. Where, in the first wave of open-air museums, buildings were moved to better proclaim a local, regional or national identity, a new generation of folk museums is growing up to express not local insularity, but to identify a locality with wider processes, particularly migration and modernization. That is, buildings are deliberately relocated at great distances precisely to make the point that people too have moved great distances. Thus, the Ulster-American Folk Park outside Omagh in County Tyrone (Northern Ireland) and the Museum of American Frontier Culture outside Staunton in Virginia, both present buildings that have travelled considerable distances. The Ulster-based version relies upon relocating old representative structures to land by the Mellon family farmstead, combining shop fronts from around Ulster into a pastiche of an Ulster emigration port, and creating replicas of structures associated with a typical Delaware River port and then the Appalachian frontier, structures ranging from shacks to opulent farmhouses. Its sister museum in Virginia presents a German farm from the Palatinate, an English farm from Worcestershire, and a farmhouse from Ulster, alongside a relocated but essentially local Shenandoah Valley 'American' farmhouse. These are not replicas, nor have they been brought over to show off the owners' taste or wealth. Instead they have been brought across the Atlantic ostensibly to save the structures from imminent destruction back home, but more particularly, to present to American visitors some sense of the cultures from which early Virginia frontier settlers had originally come. Britain plus German plus Ulster-Scots equals the American frontier, evident not just in school textbooks but now also in the landscape. It is possible for American visitors to stand within an actual Ulster farm which in turn stands within a series of walled fields imitating the farm's original old world surroundings. For visitors whose knowledge of England is little more than half-remembered school Shakespeare, the English farmstead conjures up a lost Elizabethan world, the world from which Virginia's English settlers presumably emerged.

In some ways this is just the traditional museum writ large. Artefacts are wrested out of context and exhibited thousands of miles away with suitable information boards, guidebooks and audio-tours. But whereas museums usually signal their strange outlandish contexts (mummies are never quite at home in Bloomsbury no matter how the backdrop suggests the Upper Nile), buildings that are authenticated both by the museum-controlled history of their trans-Atlantic movement and by their original sizes suggest not an exhibition artefact but the real thing. This is the artefact-image as both signified and signifier, an unmediated image that is no image at all, being the clearly authenticated structure, as real as the day it was built back in the old country. All the problems of representation that threaten the display of small artefacts within the traditional museum seem to be sidestepped, indeed irrelevant in the context of what has essentially been a re-creation of a heritage site, authentic in all but its coordinates of latitude and longitude, merely implied through the carefully reconstructed surroundings. Buildings' pedigrees validate more than each individual structure, confirming together the academic worth of the whole museum enterprise. A crofter's cottage at the Museum of American Frontier Culture in Virginia has not only been carefully reconstructed, but there on view is the history of the building's careful demolition, storage, and recreation. The passage of a Worcestershire farm from Norchard near Stourbridge is similarly documented and justified. The re-emergence of these various buildings in the Virginia countryside, though incongruous, is justified in terms of more than just rescuing buildings, but in terms of being central to the museum's scholarly credentials, its contribution to endeavour, a deliberate contrast with other less supposedly scrupulous building relocations such as those large Tudor mansions transferred from England to the banks of the James River.

Landscape Metaphor and Meaning

The traditional world-in-miniature open-air museum moves buildings not to highlight mobility and social change but to draw visitors' attention to a more stable world before modernization. In Northern Ireland, E. Estyn Evans (1951) assumed landscapes can be deciphered for the cultures that created them. Furthermore they could be recreated at the Ulster Folk and Transport Museum at Cultra to demonstrate the essentials that distinguish a particular heritage from superficially similar ones elsewhere, in this context meaning both Britain and the Republic of Ireland. While such analysis is often difficult it is nevertheless deemed essentially unproblematic, an assumption built into the very genre, which maintains a representational strategy based upon an unsophisticated empirical view of how we can remember the past. Evans epitomized widespread curatorial belief in the almost magical power of artefacts, large or small, to summon up the ghosts of people. However in a modernizing world besotted with historical novels, movies and recreations purporting to offer a glimpse of how it really was, a retreat by historians and geographers into a world of authenticated artefacts is perhaps a necessary corrective.

The Cultra site epitomizes the genre, presenting a synthetic display of Ulster heritage, much like Skansen's Swedishness (Hudson, 1987) or the Danishness of Frilandsmuseet (Michelsen, 1990; Uldall, 1990). Appropriate buildings have been brought together from across the province preserving, sharing and, even perhaps, creating memory. Across Europe, other enthusiasts for marginalized ways of life have collected often decaying buildings and created folk parks that help shape our collective pasts, partly through their association with the scholarship of archaeology and anthropology, areas deemed increasingly scientific and supposedly value-free. Such collections of the stuff of ordinary life might appear little more than the demented kleptomania of antiquarians squirrelling away anything that comes their way. But collecting to establish a research data bank associates such endeavours not with the attic but with the great collections of science from which the very bases of life could surely be discerned, making such museum collections appear instead modern and progressive. But such collecting too often reflects an increasingly out-moded scientific method: facts could be fixed, and then the debate would move on. This is more Linnaeus than Darwin, collecting to preserve in aspic and catalogue rather than to stimulate debate, the very characteristics evident within both the created landscapes of the folk museum and in the writings of E. Estyn Evans. The buildings, once recreated as they had originally been, put an end to the need for any further debate. Instead the viewer is encouraged to experience the resultant landscape in terms of the regional geography of the Annales school: *genre de vie*, *milieu*, and *personalité* (Baker, 1999). Lest such terms imply an unscientific romanticism the visitor is constantly reminded that, here at least, such terms are sustained empirically not through flights of fancy or an artist's imagination. Furthermore, artefacts large or small reflect particular milieux within which they were originally created. The buildings, their contexts and immediate surroundings are the outward and visible signs of deep cultural processes made evident through the museum's skilled preservation and reconstruction. This places a heavy responsibility upon the choice of artefacts, again verging into the domain of religious relicts. Paradoxically, it indicates that just when the artefacts are deemed most authentic and unique they are equally deemed indicative of wider themes, metaphors for otherwise inaccessible processes. That is, the artefacts are presented not for their lauded uniqueness but as a system of signs.

The creation of any assembly of buildings to typify a culture has not been without criticism, so an increasing number of museums, such as the Ulster-American Folk Park, have sought to go beyond world-in-miniature landscapes to (re-)create a series of *tableaux* (Norman, 1990). Ulster's uniqueness is presented less in its distinctive landscape than in its role in settling and thus creating modern America, the starting point of the narrative validated by the presence of at least one self-evidently genuine in situ building, the industrialist's Mellon family farmstead, and an array of relocated Ulster buildings. An otherwise long and complex historical process, emigration from Ulster to America, becomes compressed into the physical experience of the visitor. Experiencing the Ulster countryside, then the ocean crossing, and finally the arrival in Pennsylvania, the narrative creates a total framework of meaning through the presentation of simulated realities, individually plausible though unreal. It is the coherence of the overall narrative sequence, a grand tour in miniature, which strikes the viewer as real, confirming that it was Protestants that founded modern America, challenging the more widely held association of the Catholic Irish with the American urban industrial experience.

Buildings are, however, only available for the comparatively recent past, no more than the last couple of centuries. So for earlier times, or longer narratives, other strategies have been developed, particularly the creation of replica arrays. The Ulster History Museum also in County Tyrone presents rebuilt, as new, replica buildings from Prehistory to the Plantation. Most structures from this period, such as *crannógs*, rotted away long ago leaving at best a bump in a bog, or those that remain are so few in number, such as high round towers, they almost always remain *in situ*. Full-scale replica structures suggesting a temporal narrative include a Mesolithic encampment, Neolithic houses, a water-surrounded *crannóg*, a monastic centre with a round tower, plus a motte and bailey. Giving each replica a pedigree, the museum ensures visitors appreciate that they are based on the best available archaeological and historical scholarship, and are not the least bit speculative, fanciful or whimsical. With the accompanying literature it is unlikely that visitors will confuse such brand new replicas with relocated originals. This however is not the case at Glencolumbcille on the Donegal coast, where the Folk Museum presents, to the untrained eye, a hamlet of authentic structures. Yet each vernacular building is no more than a replica of those used locally in each of three successive centuries, plus a nineteenth-century school house and a sheebeen, each equipped with the furniture, artefacts and utensils of the period. This clachan, which could make the cover of any tourist guidebook, presents a view of the Ulster past that is highly problematic, implying a self-sufficient, authentically Gaelic, truly Irish community, ignoring the area's links to Scotland and later America. Indeed such a venue may tell us more about contemporary cultural politics than about the past, a salutary warning against uncritically accepting W.G. Hoskins's famous claim that 'everything is older than it seems'.

But could such problems be avoided by the rigorous use of relocated, authenticated structures, the strategy used at Staunton's Museum of American Frontier Culture to present those traditions that moved into Virginia backcountry during the late colonial period? The museum provides an array of authenticated relocated buildings from Germany, Ulster, and England, leading to the final exhibit, an American farm from elsewhere in the Shenandoah Valley.[1] Artefacts are offered

---

1   In this the museum prides itself in following the International Committee for Archaeological Heritage Management (ICAHM) requirement that 'Presentation and information should be conceived as a popular interpretation of the current state of knowledge,

to trigger folk memories, neither recognizing inherent deformations, nor challenging visitors' preconceptions about frontier life through exposure to new historical scholarship. The reconstructions invite visitors to conflate artefacts with the events they purport to represent, as at Omagh, even while the reconstructions themselves transmit not the events themselves, impossible outside a parallel universe, but one particular representation of them. Such artefacts act as signs, and thus need the kind of decoding more usually associated with literary works or movies. If historical processes are discerned through the buildings that are deemed to have emerged from such events, such structures are being treated as traces rather than as metaphors. But such fragments, however individually large or complete, remain but signs, particularly when such structures have been demolished, transported, reconstructed and finally amalgamated with other buildings and contents in arrangements that are completely novel. When such buildings are then used to confirm an empirical link between a past situation (such as being an immigrant) and a present situation (being Irish/American) the buildings used are granted a privileged status. Such synthetic landscapes invite the visitor to see amalgams as fundamentally authentic, merely because the individual buildings are deemed authentic, as if Frankenstein's monster were real merely because all its bits came from once living bodies.

Open-air folk museums, such as Omagh and Staunton, construct narratives building by building, creating a totalizing framework that surrounds the passing visitor, a narrative that validates the notion of a staid, stable, yet culturally divergent Old World of peasant dwellings and timeless ways (peat burning in the hearth, the smell of baking), all necessary prerequisites if the American frontier experience is to appear altogether novel. In this way it seems that the buildings confirm the prejudice that only in the US could people break from what Marx called the 'dead hand of history'. Both the individual buildings and the landscapes created seem entirely innocent, even while being used to create heritage rather than investigate the past. Nevertheless it remains History, as the systematic study of the past, that is evoked as the prime imprimatur that distinguishes such sites from theme parks, such as Virginia's Old Country, with its cartoon like images of Old World Europe. For the museum, a flood of authenticated details appears to confirm that what is being experienced is historical research of the finest order, even though the central feature of historical debate is absent from the public display areas. Artefacts are self-evidently more vital and more accessible than scholarly debate over what the artefacts might actually mean. The assumption built into the museum by its use of authenticated buildings is that 'what you see is what you get'. But what emerges in the open-air museum is not a landscape revealed but one equivalent to Anderson's 'imagined communities' (Anderson, 1983).

So if such museums do not present 'the real thing' do they perhaps present something even more worthwhile, simulacra better than the real thing? Replacing the image where reality once stood is something more usually associated with the

---

and [...] be revised frequently'. ICAHM 'Presentation, Information, Reconstruction', reprinted in *Antiquity*, 67, 1993, pp. 402–405.

arts, whether highbrow, as in the case of Shakespeare's overwhelming images of Julius Caesar or Richard III, or lowbrow where Disney has interposed his Pinocchio over that of the Italian original. It seems that the 'completeness' of the relocated buildings intoxicates not just visitors but the curators into seeing simulacra as being finally what's really real.

The totality of such synthetic exhibition landscapes implies an unmediated visitor experience. Concern for the authenticity of their displays leads into a form of vulgar empiricism whereby, in seeking to side-step ideological or at least theoretical issues that might confuse or alienate their visitors, they fall prey to promoting older discredited paradigms, rural-agrarian fantasies and, in effect, heritage rather than History, ignoring the problematic nature of moving buildings. Leaving a structure in an utterly changed original location may be recognized as untenable, but relocation can be just as inappropriate, for example Paul Revere's eighteenth-century colonial house incongruously stands today amidst late-nineteenth century North End Boston. But for his fame the house would have gone long ago, as have its surroundings. Leaving such buildings *in situ* may demonstrate little more than an inflexible and unimaginative, indeed fetishistic dedication to coordinates of latitude and longitude. Or can the discontinuity between a structure and its subsequent surroundings make such a juxtaposition interesting? York Minster is built, probably deliberately, out of sync with the Roman garrison commander's headquarters beneath. To remove the earlier structure merely because it doesn't fit the later medieval use of the cityscape would be to lose the experience of seeing the commander's back door with its worn down step, now exposed within the Minster foundations. To move the commander's house to a 'suitable' site would be to lose the relationship of these remains with the Roman street network within the structure of medieval York, and would rob the Minster of its proud Christian arrogance in the face of pagan Rome. If a similarly preserved Roman structure were to emerge from the path of railway expansion but a few hundred metres away, removal to a museum might be a suitable strategy if the alternative were total destruction. Maintaining Coppergate's Viking remains *in situ* despite their discontinuity with recent developments seems appropriate, and reminds us that rescue archaeology frequently involves destruction as much as rescue, and that not all of modern York's commercial needs can be satisfied within multiples of Viking frontages.

There are, however, a number of archaeologically based sites that do address the contingent nature of what they present. The Alexander Keillor museum, adjacent to English Heritage's visitor centre barn at the Avebury stone circle, a UNESCO World Heritage site, asks the visitor not just to view the exhibits but crucially to question the conventional interpretation of all that is on display. Each artefact is used to open up a line of questions rather than to provide an answer. The welcoming figure of a Mesolithic farmer appears to the arriving visitor as the kind of ill clad slave we associate with the building of colossal monuments in the ancient world. On leaving, the astute visitor might notice that the same farmer now appears more like a vividly bedecked and tattooed freeman. Both images are totally speculative, for we know nothing about the appearance or status of those who built the huge prehistoric monuments. The contrasting attire has been presented deliberately to make the case that archaeology is not a definitive study of the past, but one that presents and attempts to test hypotheses regarding the past. But for most venues, buildings as exhibits, once rebuilt, remain locked in the explanatory paradigm dominant at the founding of the museum: or as Brian Friel (1981) put it, 'remember that [...] signals [...] are not immortal. And it can happen [...] that a civilization can be imprisoned in a [...] contour which no longer matches the landscape [...] of fact'.

Conclusion

This foray into the movement of buildings has suggested that discussion of relocation is too important to be taken for granted. Consideration of what happens once a building becomes a site, a venue, an exhibit, suggests that it is necessary to understand just what representational strategies are being utilized. Visitors and critics alike need to be aware that viewing buildings wrenched out of their context may actually inhibit our understanding of the past. With suitable recognition of the theatrical implications involved, and a willingness to thump anyone foolish enough to talk about 're-creating' or 'entering' the past, relocated buildings can be part of wider strategies that enable the public and scholars alike to understand the past a little more, and to enjoy doing so. But it behoves scholars to recognize hazards involved in using relocated buildings, and to encourage people to explore the problematic aspects of the past, not just enjoy the past so easily turned into a themed attraction. Heritage is too interesting and too much fun for that, and history just too important. Relocated buildings at least deserve a warning, just as William Morris once insisted that the restored part of buildings should not be antiqued, so that the visitor would be immediately aware that the new sections were not the past itself, nor traces even, but intelligent hypotheses about the past, nothing more. Moved buildings perhaps need a health warning: 'Beware: this relocated building can seriously challenge your view of the past.' Such a paradigm shift might not be so far fetched. Only a generation ago the stately homes of Britain, Ireland and Virginia presented only the lives of once aristocratic families. Today people expect to be able to explore upstairs just as much as downstairs. From Erddig in North Wales, to Strokestown in the Irish Midlands on to Monticello in the Virginia Piedmont, their stories now include the lives of men and women, servants and slaves. If the dispossessed can come to centre stage, so too can other concerns: history from the bottom up can now be joined by considering the past as questions rather than as facts. Relocated buildings can be used to explore not just the world of those who originally built them, but the concerns of those who relocated them.

References

Anderson, B. (1983), *Imagined Communities*, Verso, London.

Baker, A.R.H. (1999), 'Reflections on the Relations of Historical Geography and the Annales School of History', in Clark, S. (ed.), *The Annales School: Critical Assessments, Volume II*, Routledge, London, pp. 96–129.

Evans, E. Estyn (1951), *Mourne Country: Landscape And Life In South Down*, Dundalgan Press, Dundalk.

Friel, B. (1981), *Translations*, Faber, London.

Greenhalgh, P. (1988), *Ephemeral Vistas: The Exposition Universelles, Great Exhibitions and World Fairs, 1851–1939*, Manchester University Press, Manchester.

Hudson, K. (1987), *Museums of Influence*, Cambridge University Press, Cambridge.

Jordan, T.G. and Kaups, M. (1989), *The American Backwoods Frontier: An Ethnic and Ecological Interpretation*, Johns Hopkins Press, Baltimore.

Michelsen, P. (1990), *Fortid Paa Fridland: Hvordan et Ffrilandsmuseum Blive Til*, Greens Forlag, Copenhagen.

Mills, S.F. (1996), 'The Presentation of Foreigners in the Land of Immigrants: Paradox and Stereotype at the Chicago World Exposition', in Materassi, M. and Santos, M. (eds), *The American Columbiad: Discovering America, Inventing the United States*, VUP, Amsterdam, pp. 251–265.

Mills, S.F. (2003), 'Open Air Museums and the Tourist Gaze', in Crouch, D. and Lübbren, N. (eds) *Visual Culture and Tourism*, Berg, Oxford, pp. 75–90.

Norman, P. (1990), *The Eighties, the Age of Parody*, Hamish Hamilton, London.

Relph, E.C. (1976), *Place and Placelessness*, Routledge, London.

Uldall, K. (1990), *Frilandsmusee*, National Museum, Copenhagen.

Chapter 9

Military Heritage, Identity and Development: A Case Study of Elvas, Portugal

João Luís Jesus Fernandes and Paulo Carvalho

Introduction

Geographers have a very particular way of looking at place as they observe, analyze, synthesize, try to understand, search for an interpretation, and weave a narrative fabric to explain the phenomena that have, directly or indirectly, helped to model the surface of the land. Among the most important concerns of these special 'observers' of the planet, in fact, are the wrinkles on the terrestrial surface, whether these are in the physical environment or whether they are of a cultural nature, as is increasingly the case. Geographers not only reflect on the structural paths that, as some believe, could lead us to homogenization, they are also interested in differences, limits, discontinuities, barriers and boundaries. Geographical spaces are seen today as complex entities, which are multidimensional, and therefore more interesting.

The study of place involves two levels of analysis: first, places should be regarded as originating from the interaction of different geographic scales. Each one is the result of a special encounter between local and external phenomena, between various processes and actors who operate at different scales. In Portugal, for instance, it is hard to understand a place without setting it on a national scale, without looking at the Iberian Peninsula as a unit, without bearing in mind the European Union and, finally, without thinking about its position in the global system. Second, a place cannot be understood without reconstructing it in time, without considering its evolution. A place also bears the imprint of its history, marked by the spatial contexts of the past. Taking the example of Elvas, a small city on the Spanish-Portuguese frontier, this chapter seeks to illustrate how both these factors interact to produce challenges and difficulties for this urban centre today.

Development, Territory, Identity and Frontier

The case study we are presenting here concerns the development dynamics of a place that has very specific locational attributes – situated next to an internal EU frontier. The redevelopment strategy of this town, Elvas, has emphasized the idea of promoting a unique identity through difference and a particular patrimonial legacy.

The case study is an excellent illustration of the kinds of questions currently of concern to geographers. First of all, what is now at the centre of the geographical debate is how each territory, even at a micro-scale, fits into the global context. Globalization, with the flows that it implies and the horizontal and vertical interdependencies that it fortifies, has become a constant challenge for places and people who must now re-assess their place in the world (Dollfus, 1998). As Johnston (1998) and Dicken and Öberg (1996) argue, this discussion is being given greater priority, not least because the world is now showing a potential for a strong and dynamic rate of change. These dynamics have an impact on people, on structures and on economic and political authorities. They are felt in the rapid compression of space-time, and in behaviours and values, especially in the way geographical space is perceived, memories are redefined and each place and its people are involved with their past and their legacies (Harvey, 1999).

The second point discussed in this analysis is the importance of territory in the twenty-first century, a century of space-time compression, shortening distance-time and falling costs. Turning our back on the temptation of the many who are bold enough to predict the end of territories, the theoretical viewpoint expounded here reaffirms the significance of the territorial dimension of societies (Badie, 1996). Societies construct different spaces. Today they have a different relationship with the territories that they organize, but they are still territorial societies and continue to express their cultures and desires in a geographical space. De-territorialization is a myth and time-space compression has not devalued the territorial dimension of societies (Haesbaert, 2004). In fact, affirmation strategies embrace territory, places and the special relationship the latter have with other places, with the past and, despite this, with the specific peculiarities cultivated in each of them. This angle implies the redefinition of traditional concepts, such as frontier, traditionally viewed as a political boundary, a sovereign barrier between States, and therefore a 'stop line' that led to many dynamic settlements springing up. In the European Union especially, the concept of frontier has been altered (Nijkamp, 1998), but as Capella and Font Garolera (1998) have shown us, frontier realities remain the fundamentals of geographical analysis.

Within the European Union, many places whose identity stemmed from their special geographical location as crossing points within a rigid boundary system have lost significance as the frontier has become flexible or more open. The loss of centrality has played a part but today, even the traditional concepts of centrality and marginality are being questioned. As Walter Leimgruber points out, traditional centres themselves contain certain factors of marginality, just as those places that became marginalized as contexts changed, also have within them the potential to become relevant again (Leimgruber, 1994).

It is here that, in the scientific theory of Geography and related sciences, territories' identities and the value of difference are reaffirmed (Harvey, 1996). Even under globalizing processes that, according to critics, promote homogenization, we are now witnessing a blossoming of strategies aiming at revitalizing places through difference (Ferrão, 2001). It is in this context that discussions about heritage and legacy are meaningful, in a world where there is ever-stronger pressure to change.

The Emergence of the Cultural Dimension of Territory and Patrimonialization

Recent years have seen a revival of the study of the cultural dimension of territory (Carvalho, 2005). In fact, the movement known as new cultural geography includes not only studies on the material dimension of culture, characterizing the early phase of cultural geography (1890–1940), but studies on the non-material dimension of culture as well (Rosendal and Correa, 1999). Current work in cultural geography differs from that previously undertaken in one fundamental aspect, in that culture is considered a composite of the work of man over time (Claval, 2002, pp. 134).

This has brought with it new perspectives on old topics, as well as the addition of new topics (Corrêa, 1999). The cultural landscape, therefore, the oldest concept in cultural geography, tackled from a morphological point of view, is now interpreted with greater flexibility and contextuality (Cosgrove, 1999). Since many of the elements that make up the cultural landscape shaped by humans provoke the transfer of knowledge, values or symbols, landscape is the matrix of culture, and at the same time bears the marks of the culture that have shaped it (Claval, 2003). For geographers, the landscape as marker and cultural matrix is a mediating element in the transfer of values (Berque, 1984).

Besides the geographical structures, decoded by the regionalist landscape language, it is essential to consider another level of perception (or subjective dimensions of landscape): the lived (Frémont, 1976). This is significant because space is experienced and appraised differently by various cultures. It is also necessary to consider a symbolic dimension because landscapes are the carriers of significances, expressing values, beliefs, myths and utopias (Cosgrove, 1978).

In this context of attempts to reassert the cultural dimension of particular territories, patrimony/heritage stands out as a major theme. Patrimony/heritage is a multi-dimensional concept, marked by a growing chronological and typological flexibility. Having overcome the monumentalist view of heritage (which is also a static view), there is insistence on the need to adopt an all-embracing perspective related to the action of humans and of nature, together with an emphasis on the original and symbolic context of the heritage (Amirou, 2000).

Binding heritage to policies of development and valorization of territories is linked to the idea that these are important resources for creating renewed images of them (Dewailly, 1998), as previously discussed in this section by Moore, in relation to Dublin. Heritage is also an important resource for rebuilding identities and stimulating investment, as well as for potentially enhancing the self-esteem of the local people (Ashworth, 1994; Butler, 1998). It is also considered an essential element in individual and social well-being and people's quality of life.

In the past few years, a number of initiatives across Europe have been introduced to promote heritage in both rural and urban contexts. In the latter, there have been

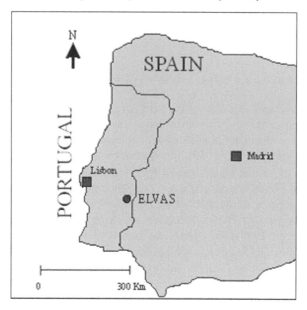

Map 9.1   The Location of Elvas

macro-interventions in the sphere of built heritage. Recently development strategies have also been increasingly linked to cultural tourism, attempts to preserve and manipulate memories, heighten identities and bolster the position of particular territories in an attempt to place them in the new global order. This has become particularly evident in our case study town, Elvas, in the south east of Portugal.

Elvas: A Case Study

Elvas is a small city in Portugal's urban network, with a population of around 15,000 in 2001. It is located on the frontier between Spain and Portugal, one of the most stable land borders in Europe (Map 9.1). This is a 1,215 km line that, apart from a few sporadic incidents, has remained untouched since the Treaty of Alcanizes, signed by the two countries in 1297. This border separates two countries with very different histories. Spain emerged from the incorporation of a series of kingdoms on the peninsular periphery, while Portugal has enjoyed its independence since the recognition of the Condado Portucalense by Pope Alexander III, in 1179. Mistrust of its neighbour, and the military infrastructure required to safeguard the country's sovereignty, meant that Portugal's land frontier achieved an important strategic value. The government adopted a strategy of planned settlement, often by banishing convicts, and built up fortified settlements along the line of the border which stand as a record in the landscape of the historical relationship between the two countries (Fernandes and Carvalho, 1998).

Map 9.2   The Historic Centre of Elvas

Elvas has had the official status of city since 1513 and a census conducted in 1527 ranked it the fifth largest city in Portugal, with a population of 8,000 people. This figure is, above all, a reflection of the strategic value of the city. Situated on a line of penetration into the country by hostile forces, Elvas assumed an important function very early on as a national defensive and protective barrier in the event of overland incursions from Spain. This fact has marked both the history of this urban centre and its identity, both of which are reflected in the urban landscape. The principles of defence and protection are essential to understanding the geography of this urban area (Map 9.2).

One of the most characteristic features of Elvas is its historic centre, a dense, compact urban landscape contained by three belts of walls: the Moorish (eighth century); the medieval (thirteenth century), and the seventeenth century wall (partly built over another section of fourteenth century walls), which enclosed the city for hundreds of years. The concentration of buildings within the walled section was first justified for defensive reasons. Nevertheless, this enclosure remained in place for a long time, even in times of peace, and it was only in the middle of the twentieth century that urban settlement spilled out beyond the seventeenth century fortifications. The symbolic value of the walled area is essential to the identity of Elvas and even today local residents refer to the location of particular places as either *intra-* or *extra-muros* (within or outside the walls).

As a stronghold during the Wars of Restoration, Elvas reinforced its defensive position and its built form reflects this function. The walls built immediately after 1640 played an extremely important role as they formed a star-shaped, bastioned perimeter of around 500,000 square metres. Thick double walls were separated by a defensive moat and opened to the outside by means of a set of double doors: narrow, hard to enter, easy to guard and securable in case of enemy attack. The defensive model of the city culminated in the construction of a number of fortresses and small forts on the periphery of the city. Of these the Forte da Graça (built in the eighteenth century), and the Forte de Stª Luzia (seventeenth century) are worthy of note. These two military edifices occupied hills overlooking the city, which could otherwise have been used as sites for enemy attack (Fernandes and Carvalho, 1998). Another feature of the city is the Amoreira aqueduct. In common with all fortresses, Elvas suffered a problem with water supply and an aqueduct was built between 1498 and 1622. It is 7,054 metres long and 31 metres at its highest point, reflecting the strategic importance of the city at that time. The military and defensive nature of Elvas is also manifest in its coat of arms that depict a knight, Gil Fernandes, a hero from the struggles against the Castilians. Legend tells us that, after a Spanish attack, he led an incursion through enemy territory to retrieve the Portuguese standard that also appears in the municipal emblem.

This strategic position influenced the daily life of the city up to the end of the twentieth century. Indeed, despite the changes in Portugal's relations with Spain, the city's military tradition was maintained until the time that Portugal's armed forces embarked on an economic rationalization process whereby they relinquished some of their military infrastructure. This rationalization occurred largely after the entry of the two countries into what was then the European Economic Community in 1986, but it was most apparent in the 1990s, when the Portuguese government began to cut public expenditure. This period saw a strengthening of economic relations with Spain, and the acknowledgement that the state of military alert between the two countries was now greatly diminished. Most of the soldiers garrisoned in Elvas left the city at that time and today many of the military buildings are disused and abandoned. This departure was significant for a city whose urban fabric is so strongly marked by its military function, evident in buildings built specifically for this purpose, and in others expropriated from religious bodies and then converted for military use. Former military quarters, barracks, the military hospital, the military court and the building that houses the controller general's department bear witness to the military legacy of this city.

Having said that, Elvas also has a substantial religious heritage. The city was the seat of a diocese in the sixteenth century, and many buildings either still belong to the Church or have been incorporated into the military structure. Although Elvas is in a part of the country where religion tends not to be overtly public, this apparent religiosity is not surprising. It could be argued that the supernatural has greater appeal at times of increased risk, and thus the interests of military and religious orders were firmly established contemporaneously in the city. The religious sites should also be viewed in the context of Elvas' experience as not only a defensive, but also a protected, city. For example, 14 of the 38 properties in the municipality of Elvas listed as National Monuments or Buildings of Public Interest can be described as military or religious architecture, while only five are listed in the category of civil architecture.[1] In an historical context of such insecurity, the chief urban agents were bound to be institutions: the State, in the shape of the military, and the Church.

But today the context is different and the city has undergone significant modifications. Although its strategic importance has long since waned, Elvas has only recently felt the consequences of demilitarization. Although the political frontier separating Portugal and Spain lost its military relevance, it did gain commercial importance, at least in the case of Elvas. Located on the main route linking Lisbon and Madrid, Elvas became one of the most important gateways between Spain and Portugal. A point of passage and a place to stop, the city lost its defensive role and embraced that of trade. The historic centre became active, not as an impregnable fortress, but as a commercial centre, open to Spanish consumers. In particular, the banking sector developed during this period taking advantage of the cross-currency trading between the escudo and peseta. Like all frontier towns, Elvas too nurtured a number of illegal activities: smuggling, and illicit currency dealings. Spain, the former enemy, because of its economic strength was now an important trading partner.

However this border role for Elvas did not last long as accession to the EEC, in 1986, strengthened the bonds between the two countries. In 1992, by now in the context of the Economic and Monetary Union, the frontiers between the two countries were eliminated. As both also adopted the single currency, the city's banking role was diminished and shortly afterwards in the 1990s, the construction of a new direct road between Lisbon and Spain, bypassing the city, was opened. Now, traffic passes around Elvas on a motorway that does not make it easy to stop. The city has lost its centrality and the 'frontier' effect, which for so many years gave the city it's character, has been weakened to the point of disappearance. Elvas now faces many questions concerning its future. Although it saw a measure of demographic recovery towards the end of the twentieth century, the city has been losing ground in Portugal's urban network rankings. From an historical ranking of fifth place in the sixteenth century, there are now almost 100 other Portuguese towns larger than it. This shift is typical of what has happened in the national, and even the peninsular, context. Portuguese settlement has now become almost entirely based along the coast. It is organized around two poles (the Metropolitan Areas of Lisbon and Porto) resulting in the intensification of socio-economic disparities between the core and periphery.

In Portugal today, the area along the land border with Spain is at the periphery of a country that has concentrated most development and opportunities along its coastline, leading some authors to declare it a 'frontier of underdevelopment' (Pintado and Barrenechea, 1974). This land border now represents the periphery

---

1   IPPAR (Instituto Português do Património Arquitectónico). *Património Classificado: Concelho de Elvas* (www.ippar.pt, 27/11/02).

of a country that, on a European scale and until quite recently, has itself also been peripheral. As the geographical and strategic importance of Elvas has declined, the city has attempted to map a future for itself and considered several possible courses. The possibilities are many because the framework or geo-political context is now more complex. The Iberian Peninsula is part of an open, globalized space, one that is competitive and rapidly changing. So how can an historic place like Elvas respond to this new reality?

The Future of Elvas

One of the possibilities for developing Elvas is to build a future that draws heavily on, and builds upon, the built heritage and memory of the past. At the end of the 1990s with this in mind, the possibility of applying for UNESCO World Heritage Site classification as a historical defensive city of global relevance was considered but dismissed.

Elvas also became a member of the Walled Towns Friendship Circle, an international network of 134 historical fortified cities. Set up in 1989, this network aims to promote sustainable development in these urban zones. Tourism, the environment and exchanging information are some of the means of co-operation. The former is one possibility that has been seriously considered and as a result the city joined the S. Mamede Tourism Region which helps promote its history, and market its built form, spectacular walls and former defensive function.

One of the forts (Sta Luzia) has been restored and now houses the Military Museum. Opened in November 2001, this museum highlights the military importance of the city. Between its opening and August 2002, about 9,000 people have visited (Câmara Municipal de Elvas, 2002). At present, the former military hospital is preparing to open as a hotel, built with foreign capital. Ironically, a building that was created as part of the defence against Spain is now being appropriated by Spanish investment. Given the wealth of megalithic archaeological heritage nearby, the *Projecto de recuperação e valorização das antas de Elvas* (Project to restore and improve the Elvas dolmens) (sponsored by IPPAR) has been gathering force since 1997, and two tourist routes taking in 18 listed megalithic monuments are being designed (IPPAR, 2000). Investing in higher education is another possibility for economic development. A college of agricultural studies has been established, which, it is hoped, will occupy some of the vacant military buildings. Industrial activity has also been promoted and now employs 17 per cent of the working population. The situation of Elvas on the Spanish-Portuguese interface will be promoted as a positive attribute in further commercial development.

However, much more difficult and one of the major challenges facing the local authorities is how to deal with the preservation and interpretation of the memory and heritage of a bygone military era. This has been a long-standing problem and one of the oldest clusters of military buildings, abandoned early in the twentieth century, had been used to house socially deprived people. Since these inhabitants have now been re-housed outside the city walls, local craftsmen are currently using the former barracks as workshops. There appears to be no overall sense of the possibilities that the built heritage can offer and a lack of direction in terms of how the town should develop.

The guidance that does exist, especially in relation to exploiting the potential of the area's history, lacks necessary supporting infrastructure. The *Piran Declaration*, a guideline document for the Walled Towns Friendship Circle, argues that 'Walled Towns are unique inheritances from times long past and should be treasured, maintained and safeguarded from neglect, damage and destruction and passed on into perpetuity as irreplaceable "Timestones of History".

But the successful manipulation of memory as a basis on which to build the future depends on a number of factors. In the first place, there is the physical dimension of this memory. Many urban monuments are abandoned, unoccupied and reoccupation has been achieved at the cost of closing monumental sites to citizens, such as the hotel, discussed earlier. The extensive fortifications that surround the town are in need of long overdue conservation works as one of the masterpieces, the Forte da Graça, is in an advanced stage of dereliction. The city's historic centre has fallen victim to completely chaotic development as buildings that have been degraded, either by inappropriate additions or by being demolished and rebuilt, have spoiled the character of the area.

A second set of problems or difficulties are created by the immaterial nature of this legacy. Local people ought to be involved in developing a future strategy as promoting the concepts of 'Identity' and 'Development' are meaningless without the backing of citizens. Educating for understanding, for intervention and for reading the landscape must begin in each local place and it is at this level that issues surrounding the changing identity of each micro-territory should be discussed. Apart from a weak and largely inactive local association for the protection of its heritage, there has been little local input into the future development of the town and its built heritage. The challenges facing Elvas demand far greater initiative and dynamism than is currently apparent. If, as we have seen, military and religious institutions once were the dominant powers in Elvas, times are different now, and a great deal more is now expected of civil society; that same civil society which, in the past, paid the taxes that paid for the building of the aqueduct, and of all the sundry military edifices (Morgado, 1992; 1993). How memory is constructed and utilized in the development strategy must be determined by means of shared initiatives, an understanding and appreciation of the past and, above all, innovation for the future.

Conclusion

In writing this chapter, we could easily have appended a sub-title such as: 'A small place in a country, in a world of many frontiers': in other words, a geographic view of a local place in an increasingly globalized world. This introduction to Elvas aimed to show that, in an interdependent, changing world, local identities also change. They are in a state of perpetual construction, through links from the past to the present and through the interaction between the local and the extra-local. The city that is the object of this study is a geographic area marked by past historical-geographical images, as well as by the interaction of scales.

Elvas appropriated a frontier identity when this borderland became a defining boundary for Portugal. But as Adriano Moreira (1997) has said, in addition to this political frontier, Portugal today has an economic frontier that does not coincide with the first; a cultural frontier that differs from both, and even a security frontier that is at odds with all the others. Elvas is no longer the custodian of one of the most important of the nation's frontiers. Today, others, too, have a strategic value. And so this small city is now trying to reassert itself as the maps are re-drawn. Rearranging the past for the present and reviving memory, in a bid to ensure a secure economic future, might be one way. For this to occur, education will be a key in rekindling and creating a collective memory, understanding and appreciation of the past among the younger generation. With their joint accession to membership of the European Union, contacts between the two sides of the frontier have intensified and there are now new opportunities for a reinterpretation and reassertion of local identity.

Globalization provides a context for new opportunities, and challenges as new maps are drawn, identities are reimagined and places struggle to retain a position for themselves in an increasingly competitive world. But in this process, memory must not be erased. Heritage and memory should not be considered a restraint, but may well be an important part of the reading of the present and, in some cases, one of the supports for the future.

References

Amirou, R. (2000), *Imaginaire du tourisme culturel*, PUF, Paris.

Ashworth, G. (1994), 'From History to Heritage. From Heritage to Identity. In Search of Concepts and Models', in Ashworth, G. and Larkham, P.J. (eds) *Building a New Heritage. Tourism, Culture and Identity in the New Europe*, Routledge, London, pp. 13–30.

Badie, B. (1996), *O Fim dos Territórios. Ensaio sobre a Desordem Internacional e sobre a Utilidade Social do Respeito*, Institut Piaget, Lisboa.

Berque, A. (1984), 'Paysage-empreinte, paysage-matrice: éléments de problemátique pour une géographie culturelle', in *L'Espace Géographique*, 23, 1, pp. 33–34.

Butler, R., Hall, C.M. and Jenkins, J. (1998), *Tourism and Recreation in Rural Areas*, Wiley, Chichester.

Capella, H. and Font Garolera, J. (1998), 'Territorial Marginalization: Regional Borders and Globalization. Some Examples from Spain', in *The Consequences of Globalization and Deregulation on Marginal and Critical Regions and Economic Systems*, IGU Commission on Dynamics of Marginal and Critical Regions Conference Proceedings, Coimbra.

Carvalho, P. (2005), *Património cultural e trajectórias de desenvolvimento em áreas de montanha. O exemplo da Serra da Lousã*, PhD thesis, Coimbra.

Claval, P. (2002), 'Campo e perspectivas da Geografia Cultural', in Rosendhal, Z. and Corrêa, R.L. (orgs) *Geografia Cultural: um século (3)*, EDUERI, Rio de Janeiro, pp. 133–196.

Claval, P. (2003), 'El enfoque cultural y las concepciones geográficas del espacio', *Boletin de Associación de Geógrafos Españoles*, 34. AGE, Madrid.

Câmara Municipal de Elvas (2002), *Boletim Municipal*, IV, 46, Câmara Municipal.

Corrêa, R.L. (1999), 'Geografia Cultural: passado e futuro', in Rosendhal, Z. and Corrêa, R.L. (orgs) *Manifestação da cultura no espaço*, EDUERI, Rio de Janeiro.

Cosgrove, D. (1999), 'Geografia Cultural do Milénio', in Rosendhal, Z. and Corrêa, R.L. (orgs) *Manifestação da cultura no espaço*, EDUERI, Rio de Janeiro.

Cosgrove, D. (1978), 'Place, Landscape and the Dialectics of Cultural Geography', *The Canadian Geographer*, 22, 1, pp. 66–78.

Dewailly, J.M. (1998), 'Images of Heritage in Rural Regions', in Butler, R., Hall, C.M. and Jenkins, J. (eds) *Tourism and Recreation in Rural Areas*, Wiley, Chichester, pp. 123–137.

Dicken, P. and Öberg, S. (1996), 'The Global Context: Europe in a World of Dynamic Economic and Population Change', *European Urban and Regional Studies*, 3, 2, pp.101–120.

Dollfus, O. (1998), *A Mundialização*, Publicações Europa-América, Lisboa.

Fernandes, J.L. and Carvalho, P. (1998), 'Heritage as a Reintegration Strategy in the Frontier Regions: An Example from the Portuguese-Spanish Border', in *Book of Abstracts of the Second International Conference on Urban Development: A Challenge for Frontier Regions*, University Press, Beer Sheva.

Ferrão, J. (2001), 'Território, última fronteira de cidadania?', in Instituto de Estudos Geográficos (eds) *Cadernos de Geografia*, Instituto de Estudos Geográficos, Coimbra.

Frémont, A. (1976), *La Région. Espace Véçu*, PUF, Paris.

Haesbaert, R. (2004), *O Mito Da Desterritorialização*, Bertrand Brasil, Rio de Janeiro.

Harvey, D. (1989), *The Condition of Postmodernity*, Blackwell, Oxford.

Harvey, D. (1996), *Justice, Nature and the Geography of Difference*, Blackwell, Oxford.

IPPAR (Instituto Português do Património Arquitectónico), (2000), *Antas de Elvas Roteiros da Arqueologia Portuguesa*, Ministério da Cultura, Lisbon.

Johnston, R.J., Taylor, P.J. and Watts, M.J. (eds) (1998), *Geographies of Global Change*, Blackwell, Oxford.

Leimgruber, W. (1994), 'Marginality and Marginal Regions: Problems of Definition', in *Marginality and Development Issues in Marginal Regions*, Proceedings of the Study Group on Development Issues in Marginal Regions, National Taiwan University, Taipei.

Moreira, A. (1997), 'Soberania de Serviço', in *Janus 97. Anuário de Relações Exteriores*, Público & Universidade Autónoma, Lisboa.

Morgado, A.O. (1992), *Aqueduto e a Água em Elvas. Fontes Antigas*, Caderno Cultural, 5, Elvas.

Morgado, A.O. (1993), *Elvas: Praça de Guerra, Arquitectura Militar*, Caderno Cultural, 7, Elvas.

Nijkamp, P. (1998), 'Moving Frontiers: A Local-Global Perspective', *Proceedings of the Second International Conference on Urban Development: A Challenge for Frontier Regions*, Beer Sheva.

Pintado, A. and Barrenechea, E. (1974), *A Raia de Portugal – a Fronteira do Subdesenvolvimento*, Afrontamento, Porto.

Rosendal, Z. and Corrêa, R.L. (orgs) (1999), *Manifestações da Cultura no Espaço*, EDUERI, Rio de Janeiro.

Chapter 10

Landscapes in the Living Memory: New Year Festivities at Angkor, Cambodia

Tim Winter

Introduction

> Landscape is a signifying system through which the social is reproduced and transformed, explored and structured. (Tilley, 1994, 34)

> In Angkor – a geographical region, an archaeological site and a cultural concept – lies much of Cambodia's future. (UNESCO, 1996, 165)

In recent years, a proliferation of studies dedicated to the understanding of national identities through the 'memories' held within symbolic landscapes have emerged (Yalouri, 2001; Boswell and Evans, 1999). This chapter uses and develops these ideas in the context of an annual four-day festival held at the World Heritage Site of Angkor, Cambodia. The legacy of Cambodia's glorious past, the templed landscape of Angkor, is revered by Khmers as a deeply symbolic icon of national, ethnic and cultural unity, all values which have been brought into sharp focus by the suffering and turmoil endured across the country in recent decades. For the purposes of this chapter, 29 interviews were conducted with individuals, couples and families at three different sites within the Angkor park: Angkor Wat, the West Mebon and at Srah Srang. Conducted in Khmer, via a translator, these semi-structured interviews focused on the values ascribed to Angkor by domestic tourists along with an examination of their activities during the four day festival. Additional insights are provided from interviews undertaken with the chief monks of two monasteries located inside the Angkor Thom complex. Analysis of the data reveals how the embodied practices of driving, picnicking, swimming and visiting pagodas are metaphorically and symbolically loaded with an optimism for the ongoing reconstruction of a nation. Accordingly, Angkor emerges as a landscape where the aspirations and visions of a future Cambodian identity are realized and articulated. In this respect, the site is not merely regarded as a cultural heritage site of the 'ancient' past, but as a form of 'living heritage'.

By considering Angkor in terms of 'memory', attention is drawn to the twin processes of forgetting and remembering recent histories, seen as vital to the (re)formation of collective identities within a post-conflict society. Indeed, an encounter with Angkor is examined as a moment when a nation's tragic past and optimistic future simultaneously intersect in an 'eternal present'. In adopting such an approach, the links between landscape temporalities and the processes of identity formation are highlighted.

Theorizing Landscape, Identity and Memory

In recent years considerable attention has been given to the fundamental dialectic between time and space within studies of landscape and place. By conceiving each as mutually constitutive, increasingly sophisticated conceptualizations have been offered regarding the often complex role landscapes play within notions of heritage and history (Boswell and Evans, 1999; Edensor, 2002). Equally, attention has also been given to the role heritage landscapes play in the formation of collective identities (Picard, 1997; Edensor, 1998).

In further developing this thread of analysis, Yalouri (2001, 17) argues that, in the case of the Acropolis, the site not only reflects certain identities but also serves to communicate and reproduce the values and meanings which underpin those identities. Accordingly, she suggests, 'the study of monument is then of necessity also a study of time and of memory ... the Acropolis [is] a "vehicle of agency" which informs the way Greeks understand their national identity'.

In contrast to earlier conceptualizations of landscape as abstract, objective and value neutral, these recent studies have centred around ideas of spatial multiplicity and the contestation arising from the variegated social actualization of place (Macnaghten and Urry, 1998; Bender, 1993; Prazniak and Dirlik, 2001). When considered together, they also represent a rich vein of academic enquiry exploring the complex ways in which the interplay between local, national and global formations of landscape and heritage intersect with the politics of ethnicity, religion and culture (Walsh, 1992; Leask and Fyall, 2000; Oliver, 2001).

By examining tourism as a form of social praxis, this chapter suggests that rather than viewing Angkor as a monumental landscape of the 'ancient' past, the site needs to be considered as a form of 'living heritage', pivotal in the articulation of cultural, ethnic and national identities. To achieve this, the chapter draws upon the recent shift towards discussing the relationship between identity, place and history in terms of memory (Connerton, 1995; Hue-Tam Ho Tai, 2001; Küchler, 2001). Rather than viewing history as held within the landscape itself, the idea of memory switches attention to the ways places and times are actively constituted and reconstituted in multiple ways on an ongoing basis (Duncan and Duncan, 1998). In this light, landscapes as *lieux de mémoire* also emerge as the medium through which multiple histories are simultaneously remembered and forgotten (Nora, 1998; McCrone, 1998).

In order to examine how this relationship between landscape and memory is articulated within the context of tourism, the recent contribution of David Crouch is helpful here. For Crouch, tourism is an encounter between people and space, but also as an encounter between material and imagined spaces. As he suggests, 'Tourism happens in spaces. That space maybe material, concrete and surround our own bodies […] [but it] may also be metaphorical and even imaginative' (Crouch, 1999, 2). Within Crouch's discussion of imaginative and metaphoric spaces we can also see an implicit concern for temporality. Indeed, the idea of metaphoric space suggests the presence of imagined pasts and futures. Conceiving tourism in such terms is central to the analysis of Khmer New Year offered here. It will be argued that constructing an account of the tourist encounter around a subject centred temporality reveals the dynamics which facilitate the formation of a series of collective identities. More specifically, it will be seen that in addition to being rendered meaningful through the presence of imagined pasts and futures, an encounter with Angkor actually serves to give meaning to an abstract Cambodian history.

Monumentalizing Angkor

The World Heritage Site of Angkor occupies around 400 sq. km. of flat plains in northwest Cambodia. The landscape incorporates four main elements: tropical forest, areas of cultivated land, a number of isolated villages, and the architectural legacy of the Angkorean period. Although assigning precise dates to 'The Angkor Period' remains a subject of debate amongst historians, it is generally recognized that the kingdom emerged as a major seat of power early in the ninth century and lasted until the capital's abandonment in the middle decades of the fifteenth century (Chandler, 1996a). Indeed, today's architectural remains testify to both the scale and wealth of Southeast Asia's greatest empire historically, covering much of what is today Thailand, Laos, Vietnam and of course Cambodia.

As the region absorbed the cultural influences of early Indian traders, a fusion occurred between Hinduism and Buddhism and the already well established indigenous forms of spirituality and religion (Chandler, 1996a). It was a synthesis which elevated Jayavarman II, popularly regarded as the first Angkorean king, into a Devaraja, or god king, on his ascension to the throne in 802. Proclaiming himself as the kingdom's first 'universal monarch', Jayavarman II was the first ruler to reign over a centrally-governed and largely unified state, one that would later become Cambodia.

Although a number of Angkorean kings built little or nothing, those who enjoyed prolonged periods of prosperity and peace typically encouraged extensive construction programmes. It was a tradition that would culminate in Jayavarman VII's vastly extravagant thirteenth-century Angkor Thom city complex. Unsurprisingly, the demands of such an extensive architectural programme are often cited by historians as a major contributory factor to the empire's eventual decline around the mid-fifteenth century (Jacques and Freeman, 1997).

The looting of Angkor by the Thais in 1432 heralded the beginning of an undistinguished period in Cambodian history and the shift of regional power towards Siam (Tarling, 1992). With temple construction superseded by a more trade-oriented society centred around Phnom Penh, Angkor's abandonment meant that the intense tropical climate and surrounding forest not only savagely eroded the stone temples but also destroyed any wooden structures abandoned by the few remaining Buddhist villagers living nearby.

Although a number of Spanish, Portuguese and Asian travellers visited the region after Angkor's demise, the late-nineteenth century travel diaries of French botanist Henri Mouhot became a pivotal moment in awakening the interest of Europeans in the site (Dagens, 1995). Encountering a labyrinth of monumental structures entangled with tree roots and lichen, Mouhot presented an account of 'discovering' Angkor in 1860 as a 'lost', even dead, civilization (Norindr, 1996). Despite the presence of numerous local villages, a powerful mythology surrounding loss and rediscovery was reinforced by the very aesthetics of Angkor's seemingly abandoned, wild and ruinous landscape, a mythology which endures today.

Notwithstanding the dubious nature of Mouhot's account, a vision of rediscovery and restitution played a crucial role in legitimizing the subsequent construction of the French administrative territory, *Indochine*. In addition, Angkor also became pivotal to constructions of a national history and identity for an emergent Cambodge (Wright, 1991). Largely through the scholarly work of the *Écolé Française d'Éxtrêmé Orient* (EFEO), Angkor was temporally and spatially fashioned as a once glorious, yet lost, cultural, national and ethnic heritage. Crucially however, Edwards points to the vital fusion between 'native and European [...] ideas of culture and politics' in the inscription of Angkor as national monument (Edwards, 1999). Indeed, as the twentieth century progressed, indigenous ideas of a noble Khmer, a Khmer cultural heritage and a Cambodian national history all converged around a totemic Angkor, and in particular Angkor Wat. Interestingly, it was 'an imagining of history and power' that continued to pervade the Khmer psyche and sense of national identity after independence was attained in 1953 (Anderson, 1991, 185). Throughout subsequent decades, imaginings of a once-glorious Angkor have remained central within Cambodia's political rhetoric, not least during the regime of the Khmer Rouge.

In April 1975, paralyzed by years of US bombing and civil war, Cambodia witnessed one of the most radical and brutal social experiments ever inflicted on a nation. Promising to liberate the country from the tyranny of both Vietnamese and American intervention, Saloth Sar, later known as Pol Pot, proclaimed the end of two thousand years of history and the return of Cambodia to 'year zero' (Ponchaud, 1978). As will be seen later, despite Pol Pot's dismissal of any historical precedents, his extreme socialist ideology was partly inspired by the once glorious agrarian civilization of Angkor, yet well over one million people, or one in seven of the population, died prior to the eventual liberation of Cambodia's capital, Phnom Penh, by Vietnamese troops in January 1979 (Barnett, 1990).

In addition to suffering the short, yet brutal, regime of the Khmer Rouge (Democratic Kampuchea), Cambodians have also endured the turmoil of an ongoing civil war in recent decades, the effects of a neighbouring US/Vietnam conflict, and occupation by the Vietnamese government throughout the 1980s. As a consequence, Cambodia has only recently begun to make significant progress towards a nationwide cultural, social and economic rejuvenation.

The international isolation of Cambodia throughout Pol Pot's regime and subsequent occupation by a Vietnamese administration, meant Angkor's conservation programme only regained momentum during the early 1990s. As Angkor formally came under the umbrella of the World Heritage Committee in December 1992, the International Coordinating Committee for the Safeguarding and Development of Angkor (ICC) was created in order to oversee efforts to protect this newly listed World Heritage Site. Incorporating all the major international and domestic organizations involved in Angkor's management, including UNESCO, the ICC oversees both monumental conservation and the development of the site for tourism. The ICC was also instrumental in the creation of the Cambodian run APSARA authority for Angkor that became operational during the mid-1990s.

Despite the various political appropriations of Angkor since Cambodia's independence, the site has also retained its widespread populist appeal and iconic status as a national, ethnic and cultural symbol. Indeed, for a population which is over 90 per cent Khmer, it is hard to overestimate the deeply symbolic significance of Angkor within Cambodia today.

Khmer New Year

In turning to consider Khmer New Year, the aim here is to identify some of the values and meanings Cambodians ascribe to a festival which has become an important, yet often overlooked, aspect of Angkor's current development as a tourist space. The analysis presented examines the New Year festival in relation to the atrocities endured across Cambodia in recent decades. Crucially, however, it becomes clear that such historical events do not simply exist as external realities for Cambodians; instead they are articulated through subjective, embodied experiences at Angkor today. In recent years, this annual event has come to symbolize a recovery from the social, political and economic forms of oppression which have characterized Cambodia's recent past.

For the majority of Cambodians the New Year serves as a welcome 'liminal, time out' from the agricultural efforts of an increasingly hot, dry season (Turner, 1995). Although not officially promoted or advertised by the government, a decade of relative political stability has enabled ever-greater numbers of Cambodians to travel to Angkor from around the country for the mid-April celebrations. Without any ticketing system in place, assessing the scale and scope of the festival over recent years remains highly problematic. However, the suffering and destruction endured during the Vietnam war and Pol Pot regime, which essentially removed any possibility of large scale celebrations, means that the volume of Cambodians visiting Angkor over the new year period today is unprecedented. In terms of calculating actual numbers, estimates of visitor numbers offered by APSARA and UNESCO range from 100–250,000.[1]

Staying in local hotels and guesthouses, tourists typically spend between two and four days visiting Angkor. In the absence of any formally organized events or celebrations, the festival is characterized by families, couples and individuals moving between a broad range of activities. In addition to visiting the numerous Angkorean temples within the park, regular visits and offerings are made to a number of modern Buddhist monasteries. Typically, hot afternoons are either spent swimming at the West Mebon reservoir, driving around the park in open top vehicles or relaxing at a number of picnic spots, the most popular of which is the west gate of Angkor Wat. With a strong emphasis placed on socializing and meeting new people, the four-day festival is defined by an interweaving of leisure, tourism and religion as visitors continually move between swimming, picnicking, temple visits and prayer.

As noted earlier, Angkor represents the material legacy of a once glorious past for Cambodians today. The interviews undertaken reveal how the site plays a pivotal role in articulating contemporary formations of cultural, national and religious identities, collective formations that have become greatly reified through the events of recent decades. Indeed, even the ostensibly innocuous activities of drinking, praying, swimming and picnicking at Angkor over new year signify a departure from the 'dark years' of the 1970s, and as such denote a national passage of time. As Meng, a local businessman states:

> It is good to see a lot of people here, and to see a lot of people employed, if there is no life at the temples then it feels like the time of Pol Pot. So I like to see many people at the temples, both visiting and working. If there is no-one selling things and no life it is like the dark years of Pol Pot. It is relief from the war to see people going to the temples and visiting them regularly. This really started from 1980 onwards, people started putting incense at the temples, so this needs to carry on today ... before we were not allowed.[2]

The sense of social liberation expressed by Meng, in terms of incense burning, remains a powerful dimension to today's New Year celebrations for a number of reasons. On 17 April 1975, the revolutionary party (angkar padevat) of Pol Pot swept into Phnom Penh declaring the end of '2000 years of Cambodian history' (Chandler, 1996a, 214). As Kiernan states 'history was to be undone, in terms of population as well as territory' by cleansing Cambodia of its religious, educational, legal and other social infrastructures (Kiernan, 1996, 27).

Despite Pol Pot's claims of returning Cambodia to 'year zero' significant elements of his radical ideology drew inspiration from a vision of a once glorious Angkorean

---

1   This lack of data was acknowledged as a significant problem during the ICC technical conference in December 2000 and again at the UNESCO/APSARA workshop on Cultural Tourism in July 2001.

2   Interview with Meng, aged 30s and a resident of Siem Reap.

period. Believing that self-sufficiency could be achieved through the planting of rubber, cotton and coconut crops, Pol Pot's goal of wealth creation via the annual export of a rice surplus was taken from a reading of Angkorean history (Vickery, 1999). Drawing on the hydraulic theories of early twentieth century French scholars, Pol Pot's massive reproduction of Angkor's irrigation technology would hold horrific consequences for the population. As Barnett (1990, 121) suggests, 'the Angkorean dream entertained by the Pol Potists, for which tens of thousands of Cambodians died as they slaved building canals, was in large part historical fantasy'.

Based on misguided beliefs that multiple, season-defying harvests could be achieved through complex irrigation systems, grossly unrealistic aims led to progressively worse annual famines. In his examination of party speeches, Chandler also indicates how Angkor was cited as an example of the power of mobilized labour and 'national grandeur which could be re-enacted in the 1970s' (Chandler, 1996b, 246). A vision of a glorious Angkorean history ensured the site remained protected and even appropriated as a political resource within the revolutionary ideology of the Khmer Rouge.

However, by removing all freedom of travel, Pol Pot denied Cambodians the opportunity to visit and experience Angkor as a landscape of collective heritage. After two decades of steady recovery, the festivities of today therefore represent a reclaiming of the site as a collective 'memory'; as a populist symbol of history enabling Cambodians to better 'understand their national identity' (Yalouri, 2001, 17). In the words of local businessman, Meng, 'Angkor Wat is a symbol and creation of Khmer culture, a symbol of national culture. That is why it is important for me, and why it is important for me to come here.' Another interviewee agreed,

> Angkor is the Khmer ancestor heritage and each year we like to see more and more people here at new year. It is a place many people want to come and it is good fortune to come as many people are still unable. After the war many people want to see Angkor, to see their heritage.[3]

Kiernan's reference to the undoing of history in territorial terms raises the importance of spatiality in this reclamation of a Cambodian 'imagined community'. Subsequent to the evacuation of the country's major urban centres in April and May 1975, the country was divided into seven zones (phumipeak), comprised of 32 administrative areas. In an effort to erase animistic traditions, abolish private land ownership, maximize human resources, as well as fundamentally reconfigure the country's political demography, major programmes of forced migration across these regions were implemented. With the majority of families dispersed across camps of forced labour, all freedom of movement was abolished. In effect, Pol Pot's revolutionary experiment represented one of the most profound severing of ties between an entire national population and its geographical base.

Needless to say, it was also a four year period straddled by further major political and social turmoil. Beginning with the American bombing campaigns across eastern

---

3   Interview with Nop, aged 30s, and living in a village 20 kms north of Angkor Thom.

provinces during the early 1970s right through to the Khmer Rouge's retreat to the provinces along the Thai border in the late 1990s, Cambodia's physical infrastructure has suffered long term destruction and neglect. Together with abject poverty, these events have meant Cambodians have been politically, economically and physically inhibited from freely travelling across their own country. The possibility of travelling to Angkor from different provinces today therefore represents an ongoing rehabilitation from this situation. As Howan, a shop owner from the northwest town of Battambang, puts it;

> Yes I like the crowds over the new year period. It's a good atmosphere meeting people from other provinces, I like to see that and I like to talk to people from other provinces because we are the same nation. I like to come and see people from different parts of the country. Since I was young, people came to Angkor Wat, up until 1968, before Lon Nol, and then it started again in 1979. But it was more local people then because it was under Vietnamese control ... since the late 1980s more and more people can come, which is good for Cambodia.[4]

Similar sentiments were expressed by Li, a woman from a family of twelve;

> We met when we came here, some of us met in Phnom Penh. We are all from different provinces now, but we are family, we are all Khmer. We are staying three days. We have heard about Angkor for a long time, and it is the first time we are all able to come together after many dark years for Cambodia. But it is Khmer heritage, built by our ancestors. We wanted to meet here.[5]

Likewise, for a family living in areas occupied by the Khmer Rouge as recently as 1998, New Year clearly represents a new era of national stability and freedom to travel;

> It's good we can now come here over new year to see the crowds, to worship in the temples. I used to come for one day on my own since 1994, but it was not safe to bring my family as we live near the Thai border ... I was a soldier on the Thai border.[6]

In these remarks we can see that the festival period represents a metaphoric rebuilding of the nation through a reclaiming of traditions, territories and material heritage. These personal touristic experiences of Angkor over New Year enable the events in Cambodia's recent history to be simultaneously remembered and forgotten. In this respect, the festival facilitates the emergence of a social memory that valuably informs the articulation of a collective identity. Indeed, by understanding New Year as a series of socio-cultural practices we are reminded of Turner's notion of *communitas* (Turner, 1995). He argues festivals represent liminal moments occurring in both time and space within which collective identities can emerge.

---

4   Interview with Howan, aged 40, travelled down from Battambang for 3 days.
5   Interview with Li, aged 50s, a woman living by the Vietnamese border in Kampong Cham.
6   Interview with Hong, aged 30, resident of Pailin.

Figure 10.1   Afternoon Picnicking at Angkor Wat
Source: Tim Winter

Meeting people and developing friendships is an experience that depends on the dynamics and symbolic potency of Angkor's landscape. Indeed, within Crouch's account of tourism as practice, he suggests friendship 'is embodied because it ... makes particular use of space. People being physically together, sharing activities, the body becomes aware of a shared body space that is also social space' (Crouch, 1999, 272). Crucially however, it is only over the time of new year when this social space is fully constituted through the collective picnicking, praying, swimming and socializing of thousands of Cambodian visitors. It can thus be suggested that this festival represents a unique time/space moment of communitas within which a sense of a nation in socio-economic and cultural recovery is collectively articulated. This is further illustrated in the following responses, from visitors based in different parts of the country:

> We come every year to see Angkor Wat, to have fun with the family, to see lots of people. We just drive around. A few families from our village have come. We've been coming for the past five years and now it's different in that there are a lot more people here and there has been more restoration. We don't know much about the history, so we like to picnic here [Angkor Wat] and the baray [reservoir]. These places have the most people, the other places are too quiet. We like to see it crowded, both with foreign people and Cambodian people. No matter how busy we are, we come here over new year, we feel we have to come to see the people at this time.[7]

---

7   Interview with Hok and Huant, aged 30s, travelled up from Kampong Thom province.

Figure 10.2   Swimming at West Baray
Source: Tim Winter

> We like to picnic here at the waterwheel, at the Baray or at Angkor Wat. It's cooler, close
> to the water, good for picnics and it's a good atmosphere at those places. Over new year
> we like to feed the monks at the Bayon, go to Phnom Kulen to see the waterfall and people
> there and worship the spirits of Khmer mythology at Banteay Srei, Neak Pean, Preah
> Khan. I feel it is very important for my family to see this, to do these things and be here
> at this time.[8]

In architectural terms, Angkor is symbolically charged not only through its Angkorean
temples, but also through the presence of numerous Buddhist pagodas constructed
within the last fifteen years. Over the new year period, frequent visits to pagodas
centre around prayer, blessings and the personal merit attained from feeding monks
and financially supporting the ongoing construction of the buildings. The following
responses given by two head monks of pagodas situated within the Angkor park
vividly illustrate the important role their monasteries play during the festival:

> There are many people that come here, they come from towns, from villages and they
> go to many monasteries, not just this one. They want to pray here because the monks are
> here. They use the old temples in the same way as the pagodas, using incense, praying to
> the Buddha. Many people like to pray for their relations so they bring food to the monks.

---

8   Interview with Sok, aged 20s, from Siem Reap province.

They believe food earns them merit as when they give food to the monk he prays for their relative so the food feeds the spirit.... People have given money to help build the Vihear, it brings them good luck and when that happens they believe in that pagoda and keep coming.

Similarly:

People come from Siem Reap town, the province, Battambang, and from the border with Thailand. They especially like to come during the festivals. They like to come to Angkor during the festival time, they like to give money. They believe all the Wats inside the Angkor area are more significant than Wats in their local areas, so they like to give donations.[9]

Clearly, the potency of these pagodas is derived not through a sense of antiquity, but through a series of value regimes very contemporary in nature. More specifically, they are valued because they are active, living sites. This ongoing support of the monastic communities and their architectural environment suggests Angkor's landscape needs to be conceived very much as a 'living heritage' rather than merely a cultural heritage of the 'ancient' past. Moreover, as many of these pagodas remain in the process of being rebuilt after their neglect and destruction during the 1970s, this sense of recovery is a far from complete project. In a similar fashion, underpinning the various responses presented here is optimism that many of the temples from the Angkorean Period will also be significantly repaired over the coming years. Chhin who comes from a small village north of Angkor and Sok from Siem Reap both agree:

Yes I would like to see the temples restored because the country has suffered four generations of war and so a lot of Cambodian people aren't knowledgeable about the Khmer history. I was born in Siem Reap. I used to come to the temples during the Sihanouk period and I used to come to the temples until Pol Pot, when I had to leave. I had to move 200 kilometres away, but I returned here in July 1979, I came back to Angkor. I am happy to see it being restored, especially the Baphuon temple. Philip Groslier was restoring it up until the war and now they are doing it again, which I am happy to see. People in the country are so poor and for them to see the glories of these temples restored makes them happy, to see Cambodia's glory restored once again, yes it makes people very happy.[10]

Yes I want to see Angkor as a modern tourist site, but they need to keep the traditional structures. They should restore the outer moat of Angkor Wat for Cambodian people to sit on, and restore these ruined temples for Cambodians to see their heritage. Cambodia is now at peace and to see Angkor restored is good for the country.[11]

---

9   Interview with head monks of pagodas situated within the Angkor park.

10 Interview with Chhin, aged 50s, who lives in a village 15 miles north of Angkor, and was visiting for the day.

11 Interview with Sok, aged 20s, a farmer from Siem Reap province who arrived by truck, and was staying at Angkor for 4 days.

Similarly, Hong who is also visiting for a number of days believes:

> Yes, to see more restoration is good, it would be nice for foreigners and Cambodians to
> see Angkor Wat and all the other temples restored. Cambodians need to be proud of their
> heritage and country and for many it is only now that they are able to come here. It is
> important for them to see Angkor rebuilt, it gives strength to our poor country.[12]

Within these various responses we can see that time spent at Angkor draws
Cambodians into the future. It is also a vision of the future defined by an optimism
that ever greater numbers of Cambodians will be able to share in the ongoing
restoration of Angkor's ancient and contemporary landscape. This feeling is shared
irrespective of age, geographical or social boundaries:

> Angkor Wat is for the young generations, it is good to see the crowds coming, it is good
> for the younger generation to see more and more people coming to the temples. I believe
> they will restore it one day. Angkor Wat is different from other countries in terms of the
> architecture. It gives strength to the Khmer people.[13]

Whether it be US bombing campaigns, neighbouring incursions or domestic
genocidal regimes, Cambodia's recent decades have been dominated by widespread
social, political and economic oppression. The interviews presented here vividly
indicate how the New Year celebrations symbolize a collective resistance and will
to recover from that period. Angkor represents a convergence of past histories, both
glorious and tragic, erased, remembered and transposed into optimistic visions of
the future. However, we have also seen how such temporalities of history don't
just exist as external realities for Cambodians, rather they are articulated through
personal, embodied experiences of Angkor today. In this respect Angkor serves
as a metaphoric space for a nation in recovery. Not only is it a recovery of the
geographical and ethnic presence of a population, it is also a recovery of their past,
present and future.

Conclusion

As a landscape of material culture built between the ninth and fifteenth centuries,
the temples of Angkor remind Cambodians of a collective former glory. Yet, the
accounts of Khmer New Year presented here have highlighted how the site is also
valued as a contemporary, socialized landscape. Examining tourism through the
lens of spatial practice has powerfully illuminated the ways in which activities such
as picnicking, swimming and driving are symbolically and metaphorically imbued
with a sense of socio-cultural recovery. Drawing upon the concept of memory has
illuminated how the recent traumatic events of a nation are simultaneously re-

---

12 Interview with Hong, aged 30, a resident of Pailin, and staying in Angkor for 3 days.
13 Interview with Suoan, aged 20s, a businessman from Phnom Penh, in Siem Reap
staying for 4 days.

Figure 10.3 New Year Celebratory Truck Driving
Source: Tim Winter

appropriated, remembered and forgotten through the personal experiences of being a tourist at Angkor today. Furthermore, we have seen how the ongoing reconstruction of Angkor's temples and modern pagodas, along with a sense of an increasingly popular festival, also provides Cambodians with an optimism for the country's future.

Addressing the complex interplay between time and space in this way provides valuable insight into the processes by which encounters with heritage landscapes can be translated and abstracted into formations of collective identities. Clearly, new year at Angkor is an example of a tourism practice whereby 'the activity and its space are enlarged in the imagination' (Crouch, 1999, 271).

As the case study of post-conflict Cambodia vividly demonstrates, we need to appreciate how the material heritage of 'ancient' monumental landscapes does not merely remain part of a nation's past, but can also actively serve as a 'living heritage' contributing to the ongoing constitution of national, cultural and ethnic identities. Unfortunately, such understandings are often ignored within heritage management policies that often conceive sites such as Angkor as 'dead' landscapes of the ancient past. In addition, this chapter has also demonstrated the importance of examining how these landscape-identity relationships emerge from, and are fashioned by, broader socio-political dynamics.

Acknowledgements

I would like to thank Mr Khin Po Thai for his extensive assistance in translating the interviews presented here; the British Academy for generously funding a period of post-doctoral fieldwork in Cambodia; and the British Sociological Association for providing a travel grant for the presentation of this paper in Dublin at the IGU Cultural Group Meeting – Perspectives on Landscape, Memory and Identity, 12th–14th December 2002.

References

Anderson, B. (1991), *Imagined Communities: Reflections on the Origin and Spread of Nationalism*, Cornell University Press, London.

Barnett, A. (1990), 'Cambodia Will Never Disappear', *New Left Review*, 180, pp. 101–125.

Bender, B. (1993), *Landscape, Politics and Perspectives*, Berg, Oxford.

Boswell, D. and Evans, J. (eds) (1999), *Representing the Nation: A Reader, Histories, Heritage and Museums*, Routledge, London.

Chandler, D. (1996a), *A History of Cambodia*, Westview Press, Colorado.

Chandler, D. (1996b), *Facing the Cambodian Past: Selected Essays, 1971–1994*, Silkworm Books, Chiang Mai.

Connerton, P. (1995), *How Societies Remember*, Cambridge University Press, Cambridge.

Crouch, D. (ed.) (1999), *Leisure/Tourism Geographies: Practices and Geographical Knowledge*, Routledge, London.

Dagens, B. (1995), *Angkor: Heart of an Asian Empire*, Thames and Hudson, London.

Duncan, J. and Duncan, N. (1998), '(Re)reading the Landscape', *Environment and Planning D: Society and Space*, 6, pp. 117–126.

Edensor, T. (1998), *Tourists at the Taj: Performance and Meaning at a Symbolic Site*, Routledge, London.

Edensor, T. (2002), *National Identity, Popular Culture and Everyday Life*, Berg, Oxford.

Edwards, P. (1999), *Cambodge: The Cultivation of a Nation 1860–1945*, unpublished PhD Thesis, Monash University.

Hue-Tam Ho Tai (2001), 'Introduction: Situating Memory', in Hue-Tam Ho Tai, (ed.) *The Country of Memory: Remaking the Past in Late Socialist Vietnam*, University of California Press, Berkeley, pp. 1–17.

Jacques, C. and Freeman, M. (1997), *Angkor: Cities and Temples*, Thames & Hudson, London.

Kiernan, B. (1996), *The Pol Pot Regime: Race, Power and Genocide in Cambodia Under the Khmer Rouge, 1975–79*, Yale University Press, New Haven.

Küchler, S. (2001), 'The Place of Memory', in Forty, A. and Küchler, S. (eds), *The Art of Forgetting*, Berg, Oxford.

Leask, A. and Fyall, A. (2000), 'World Heritage Sites: Current Issues and Future Implications', in Robinson, M., Evans, N. and Long, P. (eds), *Tourism and Heritage Relationships: Global, National and Local Perspectives*, Centre for Travel and Tourism & Business Education Publishers Ltd, Sunderland, pp. 287–300.

Macnaghten, P. and Urry, J. (1998) *Contested Natures*, Sage, London.

McCrone, D. (1998), *The Sociology of Nationalism*, Routledge, London.

Nora, P. (1998) 'From Lieux de Mémoire to Realms of Memory', in Nora, P. and Kritzman, L. (eds), *Realms of Memory: The Construction of the French Past*, Columbia University Press, New York, pp. xv–xxiv.

Norindr, P. (1996), *Phantasmatic Indochina: French Colonial Ideology in Architecture, Film and Literature*, Duke University Press, London.

Oliver, P. (2001), 'Re-presenting and Representing the Vernacular: The Open Air Museum', in Al Sayyad, N. (ed.), *Consuming Tradition, Manufacturing Heritage: Global Norms and Urban Forms in the Age of Tourism*, Routledge, New York, pp. 191–239.

Picard, M. (1997), 'Cultural Tourism, Nation Building, and Regional Culture: The Making of a Balinese Identity', in Picard, M. and Wood, R. (eds), *Tourism, Ethnicity and the State in Asian and Pacific Societies*, University of Hawaii Press, Honolulu, pp. 287–300.

Ponchaud, F. (1978), *Cambodia Year Zero*, Allen Lane, London.

Prazniak, R. and Dirlik, A. (eds), *Places and Politics in an Age of Globalization*, Rowman & Littlefield, Oxford.

Tarling, N. (1992), *The Cambridge History of Southeast Asia*, Volume 1, Cambridge University Press, Cambridge.

Tilley, C. (1994), *Phenomenology of Landscape*, Berg, Oxford.

Turner, V. (1995), *The Ritual Process: Structure and Anti-Structure*, Aldine de Gruyter, New York.

UNESCO (1996), *Angkor – Past, Present and Future*, APSARA, Phnom Penh.

Vickery, M. (1999), *Cambodia 1975–1982*, Silkworm Books, Chiang Mai.

Walsh, K. (1992), *The Representation of the Past: Museums and Heritage in the Post-Modern World*, Routledge, London.

Wright, G. (1991), *The Politics of Design in French Colonial Urbanism*, University of Chicago Press, London.

Yalouri, E. (2001), *The Acropolis*, Berg, Oxford.

Index

Hogan, D. 43
Howells, R. 65
Humphreys, J. 62
Hussey, G. 96, 107

Iacocca, L. 37–8
identity
 collective 5–8
 contemporary 90–91
 formation 97, 98
 landscape-identity nexus 3–4
 local 29, 32–3, 34
imaginative and metaphorical spaces 135
'imagined communities' 116, 139
imagined history and power 136
investment *see* funding
Ireland
 Belfast Cenotaph *7*
 Cobh, Cork 43
 famine memory 55–67
 motorways 95–6
 rural landscape 8
 Ulster museums 113–14, 115
 University College, Cork (UCC) 55–7, 65
 see also Dublin, Ireland
Irish diaspora
 Manchester 71–2
 see also Irish-Americans
Irish Independent 99
Irish National League/United Irish League
 74, 75–6, 77, 80
Irish nationalism 69–70, 71–80
Irish Republican Brotherhood (IRB) 69, 71–2
Irish-American Cultural Institute (IACI) 42,
 43, 44–5, 47
Irish-Americans 42–4, 55, 63–4, 65, 69
 heritage industry 37, 39–40
 identity 39–40, 49–50, 65
 and other immigrants 37–8, 46–8, 49, 64

Jacobson, M. 50–51
James Joyce Street (Mabbot Lane), Dublin,
 Ireland 14, *15*
Jarman, N. 70, 71, 75
Johnson, N. 65, 70

Kealy, L. 103
Kean, T.H. 45
Kelly, S. and Morton, S. 37, 40, 52, 76

Kelly, T. 69, 72
Khmer New Year *see* Angkor, Cambodia
Kiberd, D. 60–61
Kiernan, B. 138, 139

landscape
 architecture 91
 and culture 9–10, 88–91, 123, 134–5
 management 8–11
 metaphor and meanings 113–18
landscape-identity nexus 3–4
landscapes
 rural 8, 88–90
 symbolic 133, 139
 urban redevelopment 95–8
Lautenberg, Senator 47
Lewis Cairns, Scottish Highlands 22–34
Lewisian gneiss 27–8, 33
lieux de mémoire 6–7, 8, 21, 134
local identity 29, 32–3, 34
Lowenthal, D. 19, 96, 97
Lukes, S. 70

Mabbot Lane (James Joyce Street), Dublin,
 Ireland 14, *15*
McBride, I. 70
McCabe, D. 75, 76, 78
McCarthy, M. 99, 100
Macdonald, M. 29
Macdonald, S.J. 40–41, 52
McIver, N. 32
MacLean, W. 27
Macleod, A. 24, 25, 27, 28, 32, 33
McManus, R. 99, 105
Maguire, P. 103–4
Malin, D. 45
Manchester Courier (MC) 72, 74, 77
Manchester Evening Chronicle (MEC) 77
Manchester Evening News (MEN) 72, 74, 80
Manchester Examiner and Times 69
Manchester Guardian (MG) 72, 75, 76, 77,
 78–9
Manchester Martyrs commemoration ritual
 69–70
 contested issues 70–71, 75–80, 81
 devising and adapting 72–5
 sources of material 72
Marx, Karl (Marxian view) 4, 116
Massey, D. 97